The twentieth century has seen biology come of age as a conceptual and quantitative science. Biochemistry, cytology, and genetics have been unified into a common framework at the molecular level. However, cellular activity and development are regulated not by the interplay of molecules alone, but by interactions of molecules organized in complex arrays, subunits, and organelles. Emphasis on organization is, therefore, of increasing importance.

So it is too, at the other end of the scale. Organismic and population biology are developing new rigor in such established and emerging disciplines as ecology, evolution, and ethology, but again the accent is on interactions between individuals, populations, and societies. Advances in comparative biochemistry and physiology have given new impetus to studies of animal and plant diversity. Microbiology has matured, with the world of viruses and procaryotes assuming a major position. New connections are being forged with other disciplines outside biology—chemistry, physics, mathematics, geology, anthropology, and psychology provide us with new theories and experimental tools while at the same time are themselves being enriched by the biologists' new insights into the world of life. The need to preserve a habitable environment for future generations should encourage increasing collaboration between diverse disciplines.

The purpose of the Modern Biology Series is to introduce the college biology student—as well as the gifted secondary student and all interested readers—both to the concepts unifying the fields within biology and to the diversity that makes each field unique.

Since the series is open-ended, it will provide a greater number and variety of topics than can be accommodated in many introductory courses. It remains the responsibility of the instructor to make his selection, to arrange it in a logical order, and to develop a framework into which the individual units can best be fitted.

New titles will be added to the present list as new fields emerge, existing fields advance, and new authors of ability and talent appear. Only thus, we feel, can we keep pace with the explosion of knowledge in Modern Biology.

James D. Ebert
Howard A. Schneiderman

Modern Biology Series

Consulting Editors

Published Titles

ECOLOGY:

The Link Between the Natural and the Social Sciences

second edition

Eugene P. Odum

University of Georgia

Holt Rinehart and Winston

London New York Sydney Toronto

A HOLT INTERNATIONAL EDITION

Library of Congress Catalog Card Number: 74–31856
ISBN 0–03–910156–8

Photolitho reprint by
W & J Mackay Limited, Chatham
from earlier impression
First printed in Great Britain April 1975

Preface

Since the first edition of this book was published in 1963, the scope of ecology has been enlarged by public demand. As mankind everywhere has become increasingly aware of environmental abuses and limitations, the scope of people's thinking and the subject matter of ecology have broadened accordingly. In academic circles, ecology was once considered to be a branch of biology—of rather secondary importance—that dealt with the relationships of organisms and environment; now it is widely viewed in terms of the study of the totality of man and environment. Granted that the word "ecology" is often misused as a synonym for "environment," popularization of the subject has had the beneficial effect of focusing attention on man as a part of, rather than apart from, his natural surroundings. In a very real sense ecology has become a major integrative discipline that links together the physical, biological, and social sciences; hence, the subtitle of this edition: *The Link Between the Natural and the Social Sciences.*

Increased attention to things ecological inevitably brings rising expectations for solutions to critical problems; ecologists are somehow supposed to save the world even though they have had little time and few funds to check out their theories on a practical scale. As would be expected, ivory tower thinkers, specialists from many fields, one-issue one-solution advocates, and special interest proponents of varying stripe have joined the ecological bandwagon. A veritable deluge of articles and books attempt to "explain" ecology and the environmental crisis to the now concerned citizen. Encouragingly, common denominators are emerging from this cacophony of effort. I believe that the very basic principles needed to begin to understand the myriad of environmental problems were covered in the first edition of this book. However, new dimensions have come to light, and we can certainly be more specific than was possible ten years ago in pinpointing the direction that application of basic ecological principles must take if man is to achieve and maintain a mutualistic relationship with nature. Updated principles and their applications are what is new in this edition.

As was the case in the first edition, this book is organized around a series of graphic or pictorial models that illustrate the principles of ecology in a manner easily understood by the beginning student and layman alike. These illustrative models take the form of comprehensive graphs, tables, and flow charts in which symbols representing properties and forces are linked together to show how components interact. Such charts are especially useful for depicting both structure and function so that relationships between them can be emphasized. Also, "picture models" are the first step in the analysis of complex situations. Students who wish to continue beyond this introductory step can logically proceed to greater detail on the one hand and more rigorous mathematical models on the other.

Two other viewpoints underlie the presentation. The concept of levels of organization is the first principle presented and one that underlies the holistic theme of this book. Secondly, man is considered to be a dependent part of ecological systems. The impact of man's fuel-powered systems on the natural sun-powered environment is viewed as an internal, rather than an external, problem. Therefore, there is no separate chapter or appendix called "man and nature"; the whole book is as much an introduction to human ecology as to general ecology.

I wish to thank the W. B. Saunders Company of Philadelphia for permission to adopt certain features used in the third edition of my textbook *Fundamentals of Ecology*. It should be emphasized that the present volume is not a watered-down version of the larger book. Although both deal with principles developed in a whole-to-the-part progression, the present book is structured and written in a different manner for a different level of readers.

I would also like to acknowledge the assistance of those who reviewed the manuscript and offered many valuable comments: R. H. Whittaker, Cornell University; Mark M. Littler, University of California, Irvine; Ivan Valiela, Boston University; Willard Van Asdall, University of Arizona; John E. Chambers, Governors State University; Lynn Speth, Ricks College; Larry T. Spencer, Plymouth State College; Irwin A. Unger, Ohio University; George K. Harrison, University of Maryland; and Donald E. Thompson, University of South Carolina.

Eugene P. Odum
University of Georgia
Athens, Georgia
January, 1975

Contents

chapter 1

The Scope of Ecology

The term *ecology* is derived from the Greek root "oikos" meaning "house," combined with the root "logy," meaning "the science of" or "the study of." Thus, literally ecology is the study of the earth's "households" including the plants, animals, microorganisms, and people that live together as interdependent components. Because ecology is concerned not only with organisms but with energy flows and material cycles on the lands, in the oceans, in the air, and in fresh waters, ecology can be viewed as "the study of the structure and function of nature"—it is understood that mankind is a part of nature. Another useful definition that reflects current emphasis is the one of the several listed in Webster's Unabridged Dictionary that reads as follows: "the

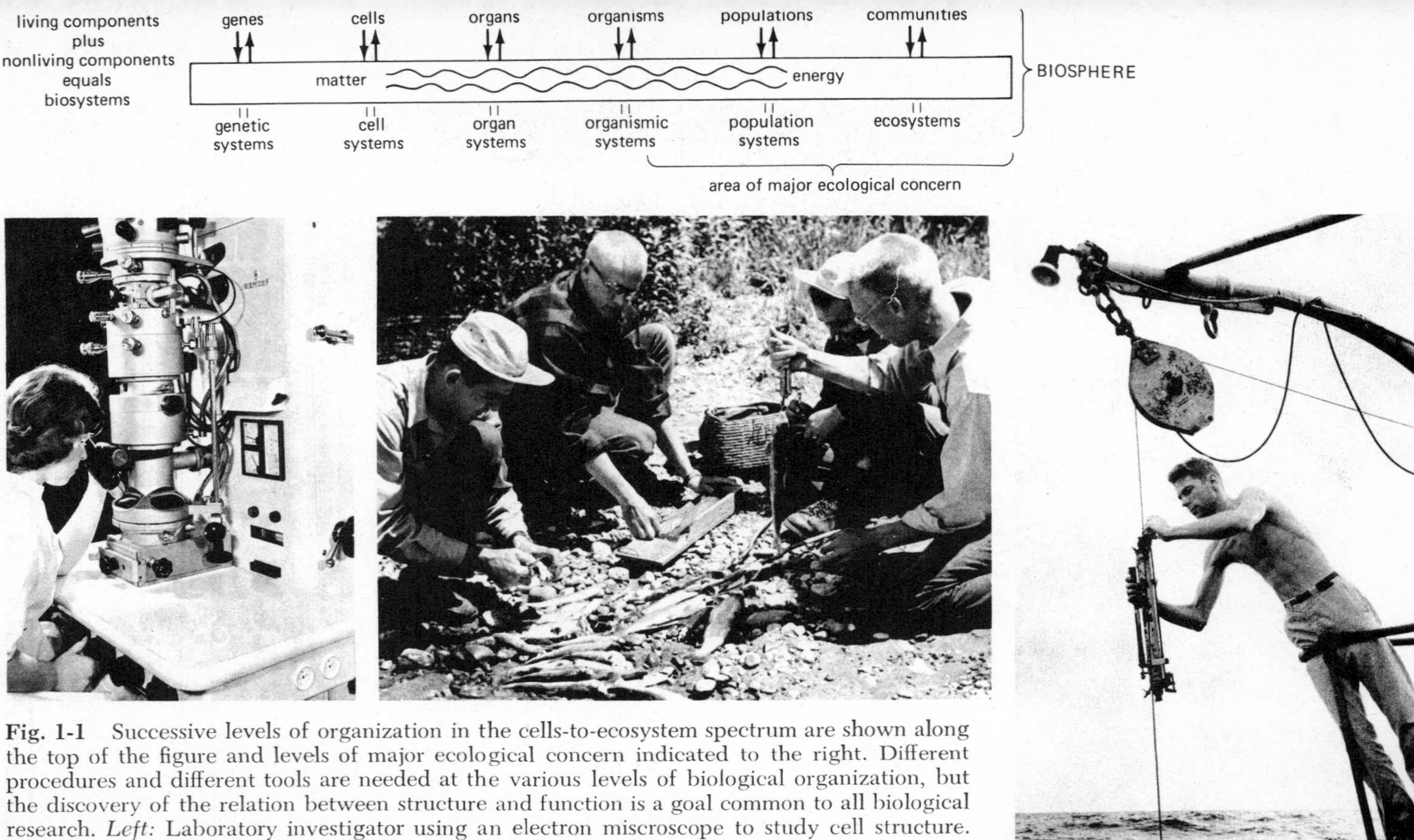

Fig. 1-1 Successive levels of organization in the cells-to-ecosystem spectrum are shown along the top of the figure and levels of major ecological concern indicated to the right. Different procedures and different tools are needed at the various levels of biological organization, but the discovery of the relation between structure and function is a goal common to all biological research. *Left:* Laboratory investigator using an electron microscope to study cell structure. (USDA photo). *Center:* Fishery ecology students studying size and composition of a population of trout. (U.S. Department of Interior Fish and Wildlife Service.) *Right:* Oceanographer lowering special sample device in the study of the sea, a natural self-sustaining solar-powered

Above: The city, a high-energy fuel-powered ecosystem that requires large areas of solar-powered ecosystems for its life support maintenance. (Dept. of Housing and Urban Development.)

totality or pattern of relations between organisms and environment."

It is interesting that the word *ecology* comes from the same root as the word *economics,* which deals with "housekeeping" in the sense of management of man's works. As will be emphasized in a later chapter in this book, extending economic cost-accounting to include the natural environment as well as man made structures and developments is an important step in redressing dangerous imbalances between these two necessary components of man's total environment. The scope of ecology has expanded considerably as man has become increasingly aware of these imbalances, an attitude change currently known as the "environmental awareness movement." Until very recently, ecology was considered in academic circles to be a branch of biology, which, along with molecular biology, genetics, developmental biology, and evolution was often, but by no means always, included in a "core curriculum" for biology majors. In this context ecology was considered to be the same thing as "environmental biology," as indeed was inferred in the first edition of this book. Now, however, the emphasis has shifted to the study of environmental systems, the whole "household" as it were, a scope that is well within the root meaning of the word, as we have seen. Thus, ecology has grown from a division of biological science to a major interdisciplinary science that links together the biological, physical, and social sciences.

Perhaps the best way to delimit the field of ecology, in terms of shifting emphasis, is to consider the concept of *levels of organization.* As shown in Figure 1-1, we may conveniently visualize a sort of *levels spectrum* in which biological units interacting with the physical environment (energy and matter) successively combine to produce a series of living systems (biosystems). The word "system" is used here in the primary dictionary sense as "a regularly interacting or interdependent group of items forming a unified whole." Ecology is concerned largely with the right-hand portion of the spectrum, as shown in Figure 1-1; that is, the levels beyond that of the individual organisms.

In ecology the term *population,* originally coined to denote a group of people, is broadened to include groups of individuals of any kind of organism. Likewise, *community* in the ecological sense (sometimes designated as *biotic community*) includes all of the populations of a given area. The community and the nonliving environment function together as an *ecological system* or *ecosystem.* A parallel term often used in the German and Russian literature is *biogeocoenosis,* which translated means "life and earth functioning together." Finally, *biosphere* is a widely used term for all of the earth's ecosystems functioning together on the global scale. Or from another viewpoint, we can think of the biosphere as being that portion of the earth in which ecosystems can operate—that is, the biologically inhabited soil, air, and water. The biosphere merges inperceptably (that is, without sharp boundaries) into the lithosphere (the rocks, sediments, mantle, and core of the earth), the hydrosphere, and the atmosphere, the other major subdivisions of our earth spaceship.

Finally, it should be emphasized that as with any spectrum, the levels-of-organization hierarchy is a continuous one; divisions are arbitrary and set for convenience and ease of communication. It is often convenient to delimit levels between those shown in Figure 1-1. For example, a "host-parasite system" which involves the interaction of two different populations would represent a level between "population" and "community."

The shift in emphasis alluded to previously has resulted from an increased interest in, and study of, the ecosystem and global levels. This does not mean that there is, or should be, any less study of organisms and populations as such. It is just that the focus in ecology has moved to the right of the Figure 1-1 spectrum. The basic reason for such a shift in emphasis stems from the realization that decisions must ultimately be made at the level of the ecosystem and biosphere if man is to avoid a major environmental crisis.

INTEGRATIVE LEVELS CONCEPT A very important corollary to the levels-of-organization concept is the *principle of integrative levels,* or, as it is also known, the *principle of hierarchical control.* Simply stated, this principle is as follows: As components combine to produce larger functional wholes in a hierarchical series, new properties emerge. Thus, as we move from organismic systems to population systems to ecosystems, new characteristics develop that were not present or not evident at the next level below. The principle of integrative levels is a more formal statement of the old adage that the "whole is more than a sum of the parts" or, as it is often stated, the "forest is more than a collection of trees." Despite the fact that this truism has been widely understood since the time of the Chinese and Greek philosophers, it tends to be overlooked in the specialization of modern science and technology that emphasizes the detailed study of smaller and smaller units on the theory that this is the only way to deal with complex matters. In the real world the truth is that although findings at any one level do aid the study of another level, they never completely explain the phenomena occurring at that level. Thus, to understand and properly manage a forest we must not only be knowledgeable about trees as populations, but we must also study the forest as an ecosystem.

In everyday life we recognize the basic difficulty in perceiving both the part and the whole. When someone is taking too narrow a view, we remark that "he or she cannot see the forest for the trees." Technologists, in particular, have often been guilty of this kind of "tunnel vision." Perhaps the major role of the ecologists in the near future is to promote the holistic approach to go along with the reductionist approach now so well entrenched in scientific methodology.

Perhaps an analogy will help clarify the concept of integrative levels. When two atoms of hydrogen combine with one atom of oxygen in a certain molecular configuration we get water (H_2O or HOH), a compound with new and completely different properties than those of its components. No matter how deeply we might study hydrogen and oxygen as separate entities we would certainly never understand water unless we also studied water. Water is an example of a compound in which the component parts become so completely bound or "integrated" that the properties of the parts are almost completely replaced by the completely different properties of the whole. There are other chemical compounds, however, in which the components partly disassociate or ionize so that the properties of the parts are not so completely submerged. Thus, when hydrogen combines with chlorine to form hydrochloric acid (HCl), the hydrogen component ionizes to a much greater extent than in water, and the properties of the hydrogen ion become evident in the acid properties of the compound. So it is with ecosystems. Some are tightly organized or integrated so that the behavior of the living components becomes greatly modified when they function together in large units. In other ecosystems biotic components remain more loosely linked and function as semiindependent entities. In the former case, we must study the whole as well as the major parts to understand the whole; in the latter case, it is easier to understand the whole by isolating and studying the parts in the traditional reductionist manner. In general, biotic systems evolving under irregular physical stress, as in a desert with uncertain rainfall, are dominated by a few species while those in benign environments, such as the moist tropics, tend to have many species with both populations and nutrients showing an intense degree of symbiosis and interdependence.

A striking example of the difference that the degree of systems integration can have on the behavior of a species component is seen in cases where insects become pests when displaced from their native ecosystems. Most agricultural pests turn out to be species that live relatively innocuous lives in their native habitat, but become troublesome when they invade, or are inadvertently introduced into, a new region or new agricultural system. Thus, many pests of American agriculture come from other continents (and vice versa), as, for example, the Mediterranean fruit fly, the Japanese beetle, and the European corn borer (the list is very long). In their original habitat these species functioned as parts of well-ordered ecosystems in which excesses in reproduction or feeding rate are controlled; in new situations that lack such controls, populations may behave like a cancer that can destroy the whole system before controls can be established. As we shall note in a later chapter, one of the prices we have to pay for high crop yields is the increasing cost of artificial chemical controls that replace the disrupted natural ones.

Some attributes, obviously, become more complex and variable as we proceed from the small to the large units of nature, but it is an often overlooked fact that rates of function may become less variable. For example, the rate of photosynthesis of a whole forest or a whole corn field may be less variable than that of the individual trees or corn plants within the communities, because when one individual or species slows down, another may speed up in a compensatory manner. More specifically, we can say that *homeostatic mechanisms*, which we may define as checks and balances (or forces and counterforces) that dampen oscillations, operate all along the line. We are all more or less familiar with homeostasis in the individual, as, for example, the regulatory mechanisms that keep body temperature in man fairly constant despite fluctuations in the environment. Regulatory mechanisms also operate at the population, community, and ecosystem level. For example, we take for granted that the carbon dioxide content of the air remains constant without realizing, perhaps, that it is the homeostatic integration of organisms and environment that maintains the steady conditions despite the large volumes of gases that continually enter and leave the air. As we shall see in Chapter 4, man's massive fuel burning may begin to overtax the capacity of nature to compensate.

The phenomena of functional integration and homeostasis means that we can begin the study of ecology at any one of the various levels without having to learn everything there is to know about adjacent levels. The challenge is to recognize the unique properties of the level selected and then to devise appropriate methods of study. In everyday language this can be restated as follows: To get good answers we must first ask the right questions. In subsequent chapters we will have occasion to cite examples of how man's progress in solving environmental problems is often slowed because the wrong question is asked, or the wrong level focused upon.

As suggested in Figure 1-1, quite different tools are needed for different levels; we do not use a microscope to study a whole ocean, a whole city, or the behavior of carbon dioxide in the atmosphere. In recent years advances in technology have expanded the scale of ecological study considerably, so that if we put our minds and money to it, appropriate measurements can be made as readily at the ecosystem level as at the individual level. Technology, of course, remains a two-edged sword. Many of man's severest problems can be traced to what might be called a "careless and arrogant," high energy-consuming technology, which runs roughshod over human values and natural laws. However, once this self-defeating and very dangerous trend is recognized, technology can be turned around to work in the opposite direction.

ABOUT MODELS In this book we will begin the discourse on ecology at the ecosystem level for reasons already indicated; it is the level of greatest interest to everybody, regardless of whether or not the subject of ecology is to be pursued beyond the introductory level. How, then, do we begin with something so formidable as an ecological system? We begin just as we would begin the study of any level—*by describing simplified versions, which encompass only the most important or basic properties and functions.* Since, in science, simplified versions of the real world are called *models,* it would be appropriate at this point to talk a little about models.

A model is a simplified formulation that mimics real-world phenomena so that complex situations can be comprehended and predictions made. In simplest form, models may be verbal or graphic, that is, consist of concise statements or picture graphs. Although for the most part we shall restrict ourselves to such "informal" models in this book, it is important that we consider at least the rationale of constructing more "formal" models, because such formulations are going to play an increasing role in decision-making regarding man's impact on his natural environment.

In its formal version a working model of an ecological situation would, in most cases, have four components, as listed below (with certain technical terms, as used by systems analysts listed in parentheses).

1. *Properties* (state variables).
2. *Forces* (forcing functions), which are outside energy sources or casual forces that drive the system.
3. *Flow pathways,* showing where energy flows or materials transfers connect properties with each other and with forces.
4. *Interactions* (interaction functions) where forces and properties interact to modify, amplify, or control flows.

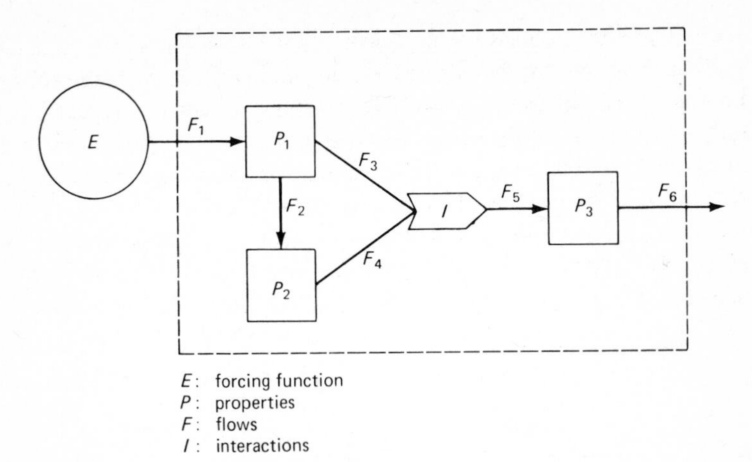

Fig. 1-2 A systems diagram showing four basic components of primary interest in modelling ecosystems. See text for explanation.

Figure 1-2 shows how these components can be linked together in a model diagram designed to mimic some real-world situation. Shown are two properties P_1 and P_2 which interact at I to produce or affect a third property P_3 when the system is driven by forcing function E. Five flow pathways are shown, with F_1 representing the input and F_5 the output for the system as a whole.

This diagram could serve as a model for photochemical smog production in the air over Los Angeles. In this case P_1 could represent hydrocarbons and P_2 nitrogen oxides, two products of automobile exhaust emission. Under the driving force of sunlight energy E, these interact to produce photochemical smog P_3. In this case the interaction function I, is a synergistic or augmentative one in that P_3 is a more serious pollutant for man than is P_1 and P_2 acting alone.

Alternatively, Figure 1-2 could represent a grassland ecosystem in which P_1 are the green plants, which convert sun energy E to food. P_2 might represent an herbivorous animal that eats plants, and P_3 an omniverous animal that can eat either the herbivores or the plants. In this case the interaction function I could represent several possibilities. It could be a "no-preference" switch if observations in the real world showed that the omnivore P_3 eats either P_1 or P_2, according to availability. Or I could be specified to be a constant percentage value if it was found that the diet of P_3 was composed of, say 80 percent plant and 20 percent animal matter irrespective of the state of P_1 and P_2. Or I could be a seasonal switch if P_3 feeds on plants during one part of the year and animals during another season. Or it could be a threshold switch if P_3 greatly prefers animal food and switches to plants only when P_2 is reduced to a low level.

These examples will suffice to show the tremendous versatility of model building, not only to provide simplified versions of the real world that help us understand it, but also to set up hypothetical test cases to answer questions about man's impact on an ecosystem as, for example, what would happen if this property were removed, or that interaction changed, or that energy source reduced? To use and experiment with models for any theoretical or practical purpose the informal chart models that we have been discussing must be converted to mathematical (that is, formal) models by quantifying properties and drawing up equations for flows and their interactions. This is a subject for advanced texts, but we think it is important that students and citizens alike understand the way in which the professional model-builder goes about his business.

Ecosystems are capable of self-development that may include internally programmed, or externally induced, growth, repair, replacement of parts, and other processes that counter the natural tendency of any and all systems to deteriorate with time. The problems of thermodynamic "disorder" and ecosystem "development" will be considered in subsequent chapters. At this point in our presentation we need only to emphasize that it is always important in modelling ecosystems to consider their behavior with respect to time. It is especially important to know whether an ecosystem is in a state of change or stable. Otherwise, how are we to anticipate future conditions and be able to judge whether a future condition is the result of a natural process or a perturbation, perhaps, man-made?

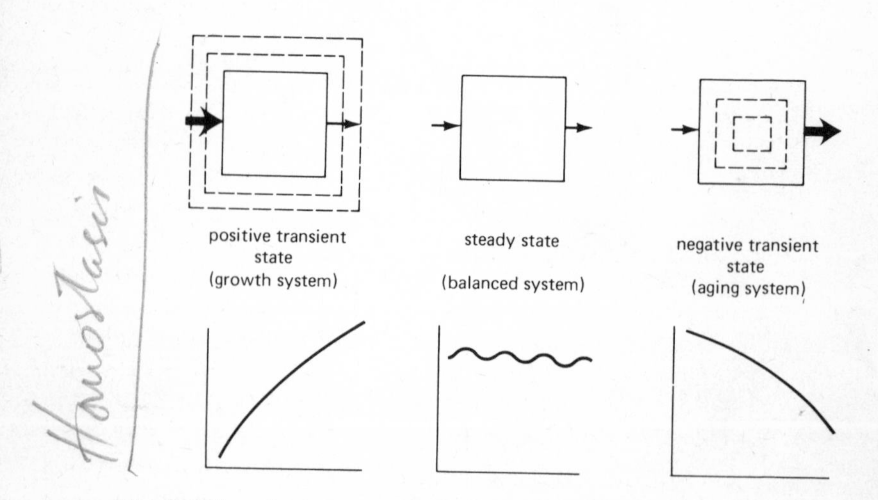

Fig. 1-3 An elementary cybernetics of ecosystems: The three "systems states" in terms of change with time.

For the moment, let us consider an ecosystem as a unit that could be represented in a diagram as a simple rectangle, or *black box*, which we can define as: "Any unit whose function can be evaluated without specifying the internal contents." Figure 1-3 shows box diagrams of three states that an ecosystem might be in with respect to anticipated change with time. Below each box is a "behavior graph" showing basic trends. The positive transient state or growth system is one in which an excess of in-come over out-go (as shown by larger input arrow as compared to the output arrow), is used to add to internal structure or make the system bigger. In contrast, a negative transient state or aging system (see right-hand box, Figure 1-3) is one in which more goes out than in, so that storage and parts are used up faster than they are replaced and the system thereby becomes smaller or less active. As shown in the middle of diagram of Figure 1-3, a steady-state system is one in which inputs and outputs are balanced. A newly planted crop or a new pond in which organisms are just beginning to colonize are examples of growth ecosystems. A fallen log would be an example of an aging ecosystem since life is sustained in the decaying log only by using up the stored energy of the wood with no new wood being added. A mature forest or an ocean which does not change in gross appearance and structure from year to year over many years can be regarded as a steady-state, since trees and other components are replaced on the average at the same rate as they die or are dispersed.

Of course, we all recognize these three states in man's special habitat. We have all seen, or perhaps lived in, towns that are growing rapidly, towns that are dying, and towns that are neither growing nor shrinking.

It should be emphasized that just because an ecosystem is in a steady-state does not mean that it is inactive. A large, mature forest like a large, mature elephant has a tremendous metabolism and requires a large flow of energy to sustain it. It seems likely that there is no such thing as a "rock steady" steady-state; fluctuations including seasonal and annual cycles can be expected. Also, diseases or storms frequently cause setbacks which are followed by periods of growth and recovery. For this reason a wavy rather than a straight horizontal line is used to graph the steady-state in Figure 1-3. Thus, there are two kinds of stability that especially concern us. Stability in time is one aspect as just discussed. The ability of a system to return to a particular state after being knocked out of equilibrium or displaced in some manner by strong perturbation (outside force) represents another kind of stability (often called "structural" or "neighborhood stability" by engineers). We will come back to this second aspect later.

We are all aware that one of the great debates of our times has to do with the question of whether (or perhaps when) man must design and adapt to steady-state conditions (with maximum ability to resist perturbations) and, at the same time, avoid the aging state. In his short history, mankind has experienced a succession of growth states with ever-increasing levels of population density and energy utilization. Consequently, even the idea that there might be "limits to growth" in man's world is a new, and, to many people, almost unthinkable consideration. If you have not lost interest in ecology by the time you have reached the end of this chapter, and if you can continue through the next several chapters where we try to establish some basic principles, then we can come back, toward the end of the book, with a summary of some of the trade-offs between rapid growth and stability, and between social justice and environmental quality that have to be considered in the future.

SUGGESTED READINGS

The scope of ecology

Hutchinson, G. E. 1964. The lacustrine microcosm revisited. *Amer. Sci.* 52:331–341. (Discusses the holological, that is, the study of the whole, and the merological, that is, the study of the parts, approaches as contrasting philosophies in the study of lakes and other complex systems.)

Platt, Robert and J. Wolfe, eds., 1964. Special issue on ecology. *Bio-Sci.* 14:7–41. (Essays on the scope of ecology as viewed by different authors.)

Integrative levels concept

Fiebleman, J.K. 1954. Theory of integrative levels. *Brit. Journ. Phil. Sci.* 5:59–66.

Prosser, C. Ladd. 1965. Levels of biological organization and their physiological significance. In *Ideas in Modern Biology*, ed. J. A. Moore, pp. 359–388. Garden City, New York: Natural History Press.

Modelling

Essentially nonmathematical discussions

Odum, Howard T. 1971. The world system. In *Environment, Power and Society*. Chapter 1, pp. 1–25. New York. Wiley-Interscience.

Van Dyne, G. M. 1966. Ecosystems, systems ecology and systems ecologists. Oak Ridge National Laboratory Report. 3957. Also In *Readings in Conservation Ecology,* ed. G. W. Cox, 2nd ed. 1974. Englewood Cliffs, New Jersey: Prentice-Hall.

Watt, Kenneth E. F. 1966. The nature of systems analysis. In *Systems Analysis in Ecology*. Chapter 1, pp. 1–14. New York: Academic Press.

For readers with good mathematical backgrounds

Dale, M. B. 1970. Systems analysis and ecology. *Ecol.* 51:2–16.

Patten, Bernard C. 1971. A primer for ecological modelling and simulation with analog and digital computers. In *Systems Analysis and Simulation in Ecology*, Vol. 1, ed. D. C. Patten. New York: Academic Press.

Smith, F. E. 1970. Analysis of ecosystems. In *Analyzing Temperate Forest Ecosystems*, ed. D. E. Reichle, pp. 7–18. Berlin-New York: Springer-Verlag.

Walter, Carl J. 1971. Systems Ecology. In *Fundamentals of Ecology,* 3rd ed., E. P. Odum, pp. 276–292. Philadelphia: Saunders.

chapter **2**

The Ecosystem

As was made clear in the previous chapter, the ecosystem is the basic functional unit with which we must deal since it includes both the organisms and the nonliving environment, each influencing the properties of the other and both necessary for maintenance of life as we have it on the earth. By considering the ecosystem first we are in a very real sense beginning our study of ecology with the gross anatomy and physiology of nature, much as a beginning medical student might begin his study with the gross anatomy and physiology of the human body. Once a clear image of overall structure and function has been obtained, component parts and functions, such as populations, the cyclic behavior of nitrogen, or plant productivity can be considered; or a particular environment, such as an ocean, a city, or a desert can be placed in perspective.

BASIC KINDS OF ECOSYSTEMS Since energy is an important common denominator in all ecosystems, whether designed by nature or by man, it provides a basis for what might be called a "first-order" classification. As will be detailed in the next chapter, energy is always a major forcing function. The source and quantity of available energy determines to greater or lesser degree the kinds and numbers of organisms, and the pattern of functional and developmental processes—not to mention the life-style of man. Therefore, knowledge about the energetics of an ecosystem is always of key importance in understanding its properties.

An energy-based classification of ecosystems is outlined in Table 2-1, together with an order-of-magnitude estimate of the range of energy utilized in terms of kilocalories that flow through a square meter on an annual basis. Since different units (joules, calories, BTU's, kilowatts, and so on) are used by specialists in dealing with different forms of energy, and since the figures in Table 2-1 may not mean much to you in the absolute sense, this might be a good time to refer to Appendix Table 1 for explanations of units and for a list of convenient conversion factors. It is important to note that some energy units, such as the watt, have time built into the definition and are thus *energy-time* or *power* units. Other units, such as the calorie, represent potential energy (not time-specific): A unit of time must be added to convert these units to power rates. In this book we shall mostly use *kilocalories* (abbreviated: kcal) per day or per year to quantify energy flow. Thus, when we speak of "power level" or use the term "powered" we are referring to energy flow per unit of time. In order to compare various kinds of ecosystems we add a unit of area such as the square meter, acre, hectare, and so on.

To relate the quantities discussed in this and the next chapter more directly to your own personal experience the following might be kept in mind: an adult person in the United States consumes about 2800 kcal of food daily, or about one million (10^6) yearly; there are 10,000 (10^4) square meters (m^2) in a hectare, 4046 in an acre. As points of reference for fuel energy, the potential energy in a pound of coal is about 3150 kcal, a pound of gasoline about 5230 kcal, a gallon of gasoline about 32,170 kcal. Now let us consider the four kinds of ecosystems as listed in Table 2-1.

Table 2-1. Ecosystems Classified According to Source and Level of Energy

	Annual Energy Flow (Power level) Kilocalories per Square Meter
1. *Unsubsidized Natural Solar-powered Ecosystems* Examples: open oceans, upland forests. These systems constitute the basic life-support module for spaceship earth.	1000–10,000 (2000)[a]
2. *Naturally Subsidized Solar-powered Ecosystems* Examples: tidal estuary, some rain forests. These are the naturally productive systems of nature that not only have high life support capacity but produce excess organic matter that may be exported to other systems or stored.	10,000–40,000 (20,000)[a]
3. *Man Subsidized Solar-powered Ecosystems* Examples: agriculture, aquaculture. These are food and fiber-producing systems supported by auxiliary fuel or other energy supplied by man.	10,000–40,000 (20,000)[a]
4. *Fuel-powered Urban-industrial Systems* Examples: cities, suburbs, industrial parks. These are man's wealth-generating (also pollution-generating) systems in which fuel replaces the sun as the chief energy source. These are dependent (i.e., parasitic) on classes 1–3 for life support and for food and fuel.	100,000–3,000,000 (2,000,000)[a]

[a] Numbers in parentheses are estimated round-figure averages, actually little more than guesses since the earth's ecosystems have yet to be inventoried in sufficient depth to calculate averages.

Ecosystems rely on two major sources of energy, the sun and chemical (or nuclear) fuels. Thus, we can conveniently distinguish between *solar-powered* and *fuel-powered* systems on the basis of the major input, while recognizing that in any given situation both sources may be utilized. It is important to note that although the total solar energy impinging upon the earth is enormous, solar radiation on an area basis is a dilute energy source, because only a small portion of that which falls on a square meter is directly usable by organisms (more about this in the next chapter). In contrast, fuel may provide a highly concentrated source in terms of conversion to useful work within a small area.

The systems of nature that depend largely or entirely on the direct rays of the sun can be designated as *unsubsidized solar-powered ecosystems* (category 1 in Table 2-1). They are unsubsidized in the sense there is little, if any, available auxiliary source of energy to enhance or supplement solar radiation. The open oceans, large tracts of upland forests and grasslands, and large deep lakes are examples of relatively unsubsidized solar-powered ecosystems. Frequently, they are subjected to other limitations as well, as, for example, a shortage of nutrients or water. Consequently, ecosystems in this broad category vary widely, but are generally low powered and have a low productivity, or capacity to do work. Organisms that populate such systems have evolved remarkable adaptations for living on, and efficiently using, scarce energy and other resources.

Although the "power density" of natural ecosystems in this first category is not very impressive, nor could such ecosystems by themselves support a high density of people, they are none the less extremely important because of their huge extent (the oceans alone cover almost 70 percent of the globe). From the human interest standpoint the aggregate of solar-powered, natural ecosystems can be thought of, and they certainly should be highly valued, as the basic life-support module which provide desirable stability and homeostatic control for spaceship earth, as mentioned in Chapter 1. It is here that large volumes of air are purified daily, water recycled, climates controlled, weather moderated, and much other useful work accomplished. A portion of man's food and fiber needs are also produced as a by-product without economic cost or management effort by man. This evaluation, of course, does not include the priceless aesthetic values inherent in a sweeping view of the ocean, or the grandeur of an unmanaged forest, or the cultural desirability of green open space.

Where auxiliary sources of energy can be utilized to augment solar radiation the power density can be raised considerably, perhaps an order of magnitude (that is, ten times), as indicated in Table 2-1.

In this frame of reference an *energy subsidy* is an auxiliary energy source that reduces the unit cost of self-maintenance of the ecosystem, and thereby increases the amount of solar energy that can be converted to organic production. In other words, solar energy is augmented by nonsolar energy freeing it for organic production. Such subsidies can be either natural or man-made (or, of course, both). For the purpose of our simplified classification we have listed *naturally subsidized* and *man-subsidized solar-powered ecosystems* as categories 2 and 3, respectively, in Table 2-1.

A coastal estuary is a good example of a natural ecosystem subsidized by the energy of tides, waves, and currents. Since the back and forth flow of water does part of the necessary work of recycling mineral nutrients and transporting food and wastes, the organisms in an estuary are able to concentrate their efforts, so to speak, on more efficient conversion of sun energy to organic matter. In a very real sense, organisms in the estuary are adapted to utilize tidal power. Consequently, estuaries tend to be more fertile than, say, an adjacent land area or pond which receives the same solar input, but does not have the benefit of the tidal and other water flow energy subsidy. Subsidies that enhance productivity can take many other forms, as for example, wind and rain in a tropical rain forest, the flowing water of a stream, or imported organic matter and nutrients received by a small lake from its watershed.

Man, of course, learned early how to modify and subsidize nature for his direct benefit, and he has become increasingly skillful in not only raising productivity, but more especially in channeling that productivity into food and fiber materials that are easily harvested, processed, and used. Agriculture (land culture) and aquaculture (water culture) are the prime examples of category 3 (Table 2-1), the *man-subsidized solar-power ecosystems*. High yields of food are maintained by large inputs of fuel (and in more primitive agriculture, human and animal labor) involved in cultivation, irrigation, fertilization, genetic selection, and pest control. Thus, tractor fuel, as well as animal or human labor, is just as much an energy input in agro-ecosystems as sunlight, and it can be measured as calories or horsepower expended, not only in the field, but also in processing and transporting food to the supermarket. As H. T. Odum (1971) has so aptly expressed it, the bread, rice, corn, and potatoes which feed the masses of people are "partly made of oil." This is why fuel, or some comparable auxillary energy, is vital to food production for man.

It is very important to note that recent increases in crop yield, the so-called "green revolution," has resulted from genetic selection of plants, not so much for their ability to utilize solar energy as for their

Naturally subsidized

Man subsidized

ability to benefit from fuel subsidies. Thus, what has been called in the popular press "miracle" rice and wheat are dwarf plants with small root systems and just enough leaves and stem to capture a maximum of usable solar radiation. Since man's fuel and chemicals do most of the work of protection and maintenance that a wild plant would have to do with an expenditure of its own energies, the crop plant is able to convert more of the sun energy into grain. It can do this because it is highly selected (that is, genetically programmed) to produce grain at the expense of nonedible tissue. Pouring the fertilizer, or other subsidies, on a wild rice plant would not have such a great effect on grain yield since the wild plant would be programmed to use the additional resources for stalks and leaves as well as grain. Man's skill in augmenting the natural conversion of sun energy into food in this fashion parallels nature's own design and has, at least temporarily, staved off starvation in some parts of the world. However, the *fuel-subsidized agro-ecosystem* is not without its economic and pollution costs resulting from the heavy energy consumption (for estimates of this, see Figure 8-2); also, the high degree of genetic specialization produces an inherent vulnerability to disease. Whether fuel-subsidized food production and rising per capita expectations can keep up with world population growth is now the question. More about this in Chapter 8.

In Table 2-1 the productivity, or power level, of natural and man-subsidized solar-powered ecosystems are listed as the same. This evaluation is based on the observation that the most productive natural ecosystems and the most productive agriculture are at about the same level; about 50,000 kcal m^{-2} yr^{-1} seems to be the upper limit for any plant-photosynthetic system in terms of continuous, long-term function. The real difference in these two classes of systems is in the distribution of the energy flow, as indicated in the previous paragraph; man works to channel as much energy as possible into food he can immediately use, while nature tends to distribute the products of photosynthesis among many species and products and to store energy as a "hedge" against bad times in what we shall later discuss as "a strategy of diversification for survival."

We now come to man's crowning achievement, the *fuel-powered ecosystem* (category 4, Table 2-1), otherwise known as the urban-industrial system. Here, highly concentrated potential energy of fuel replaces, rather than merely supplements, sun energy. As cities are now managed, solar energy is not only unused within the city itself, it becomes a costly nuisance by heating up the concrete, contributing to the generation of smog, and so on. Food, a product of solar-powered systems, is here considered to be an externality since it is largely imported from outside the city. As fuel becomes more expensive for man it is likely that interest in utilizing solar energy in cities will increase, so we can anticipate a new class of ecosystems, the "sun-subsidized, fuel-powered city" (see Figure 8-1). Also, man may find it prudent to develop a whole new technology designed to concentrate solar energy to a level where it might partially replace fuel, rather than merely supplement it. Only time will tell what should be man's best strategy for survival, but one thing seems certain; it will have to be based on a better partnership between man and nature than now exists.

As of now, two properties of the fuel-powered system need to be emphasized: First, we should take note of the enormous energy requirement of a densely populated urban-industrial area; it is at least two or three orders of magnitude greater than the energy flow that supports life in natural or semi-natural solar-powered ecosystems. As already indicated this is why many people can live together in a small space. The kilocalories of energy that annually flow through a square meter of an industrialized city are to be measured in the millions rather than thousands (Table 2-1). Thus, an acre of highly developed fuel-powered urban environment consume a billion kilocalories (about 10^9) or more each year. A more dramatic way to view this energy demand is to consider per capita consumption. In 1970 17.4×10^{15} kcal (69×10^{15} BTU) of fuel energy (including that required to generate electricity) were consumed in the United States, which, divided by 200 million people, comes to about 87 million kcal per person, per year. Recall from an earlier paragraph that only 1 million kcal per person is required for food energy. Thus, household, industrial, commercial, transportation, and other "cultural" activities in the United States use 86 times as much energy as that required for man's "physiological" needs (that is, food energy to power the body). In undeveloped countries, of course, the situation is quite different. Per capita fuel energy consumption in India and Pakistan is 1/50 and 1/100 times less, respectively, as in the United States. In such countries human and animal labor are still more important than machines, and a much larger proportion of the country's total energy flow involves food and food production.

During the past decade or so per capita energy consumption has been increasing at a much faster rate than population growth. By the time you read this, for example, annual per capita consumption in the United States will probably be well past 90 million kcal. Such a disparity is a matter of grave concern. For one thing, the rich tend to get richer faster than the poor under such an unbalanced growth pattern, which could lead to social upheavals that could bring on wars of destruction. How to achieve a better world-wide distribution of energy may be the great challenge of the next century.

The second point to emphasize is that the fuel-powered system, in contrast to natural sun-powered ones, is an incomplete or dependent ecosystem in terms of life support since it produces no food, assimilates very few wastes, and recycles only a small portion of its water and other material needs; and most of the energy that runs it comes from outside, often from great distances. Thus, an acre of a city requires not only many acres of agro-ecosystems to feed it, but even more acres of general life support, natural or seminatural environment to take care of the carbon dioxide and other large volume wastes and to supply it with hugh volumes of water and other materials. The per capita use of water, including irrigation, is something like 2000 gallons per day, of which 730 gallons are consumed (that is, not returned to streams or other sources). A city person also consumes a ton of wood products (paper, lumber, and so on) per year which requires from 0.3 to 1 acre to produce (depending on the intensity of forest management). These are just two examples of an affluent individual's impact on his environment. (See appendix 3).

To summarize, the stress that a high-powered fuel system places on the adjacent lower-powered sun system is enormous. The power differential between them increases with the power level of the city since there is a sharp upper limit to the work capacity of any system powered only by dilute sun energy. The richer the city in terms of energy use the greater the area of life support that is required, a reality city planners and developers are often strangely unaware. It is no accident that all of the world's great industrial cities are located on coasts, large estuaries, large rivers, or fertile deltas where life-support capacity of the natural environment is high, or extensive, or both. As we become more concerned with land-use planning it is important to recognize that natural, self-sustaining solar-powered ecosystems have a direct value to man for their life support and waste assimilation capacities as well as for their food, fiber, or recreational potential. Any city that overtaxes its life-support module, or fails to preserve enough of it, can find itself caught in a vicious downward spiral of declining cost-benefits as costs of paying for what was once the "free work of nature" overrides the benefits of life in the city. In Chapter 8 we will consider the urgent need to incorporate the work of nature into the economic value system so that costs and benefits can be assessed for the interdependent urban-rural complex as a whole.

HE COMPONENT PARTS OF AN ECOSYSTEM The gross structure of several different kinds of ecosystems is diagramed in Figures 2-1 and 2-2. When considered from the ecosystem point of view, a lake, a forest, or other recognizable unit of the landscape has two biotic components: an *autotrophic* component (autotrophic means "self-nourishing"), able to fix light energy and

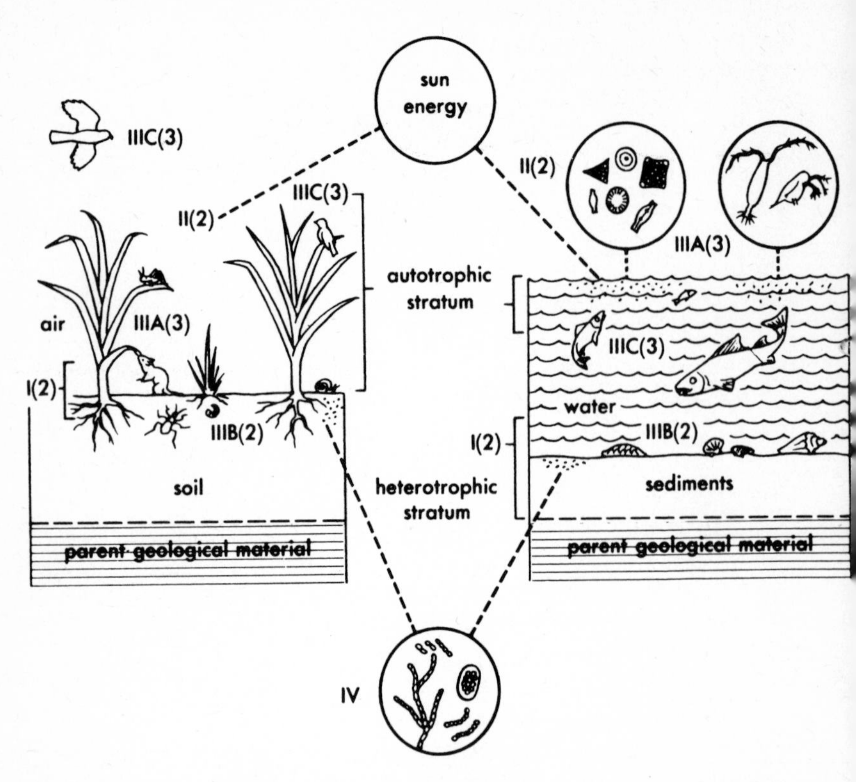

Fig. 2-1 Sun-powered autotrophic ecosystems. Comparison of the gross structure of a terrestrial grassland and an open-water ecosystem (either fresh water or marine). Necessary units for function are: I: Abiotic substances (basic inorganic and organic compounds). II: Producers (vegetation on land, phytoplankton in water). III: Macroconsumers or animals: (A) direct or grazing herbivores (grasshoppers, meadowmice, etc. on land; zooplankton, etc. in water); (B) indirect or detritus-feeding consumers or saprovores (soil invertebrates on land; bottom invertebrates in water); (C) the "top" carnivores (hawks and large fish). IV: Decomposers, bacteria and fungi of decay.

Fig. 2-2 Heterotrophic ecosystems. (A) One of nature's "cities"—an oyster reef that is dependent on the inflow of food energy from a large area of surrounding environment. (B) Industrialized city maintained by a huge inflow of fuel and food with a correspondingly large outflow of waste and heat. Energy requirement, on a square meter basis, is about 70 times that of the reef, or about 4000 kcal/day which comes to about 1½ million kcal per year (compare with Table 1-1). (After H. T. Odum, 1971. Courtesy of the author and John Wiley & Sons.)

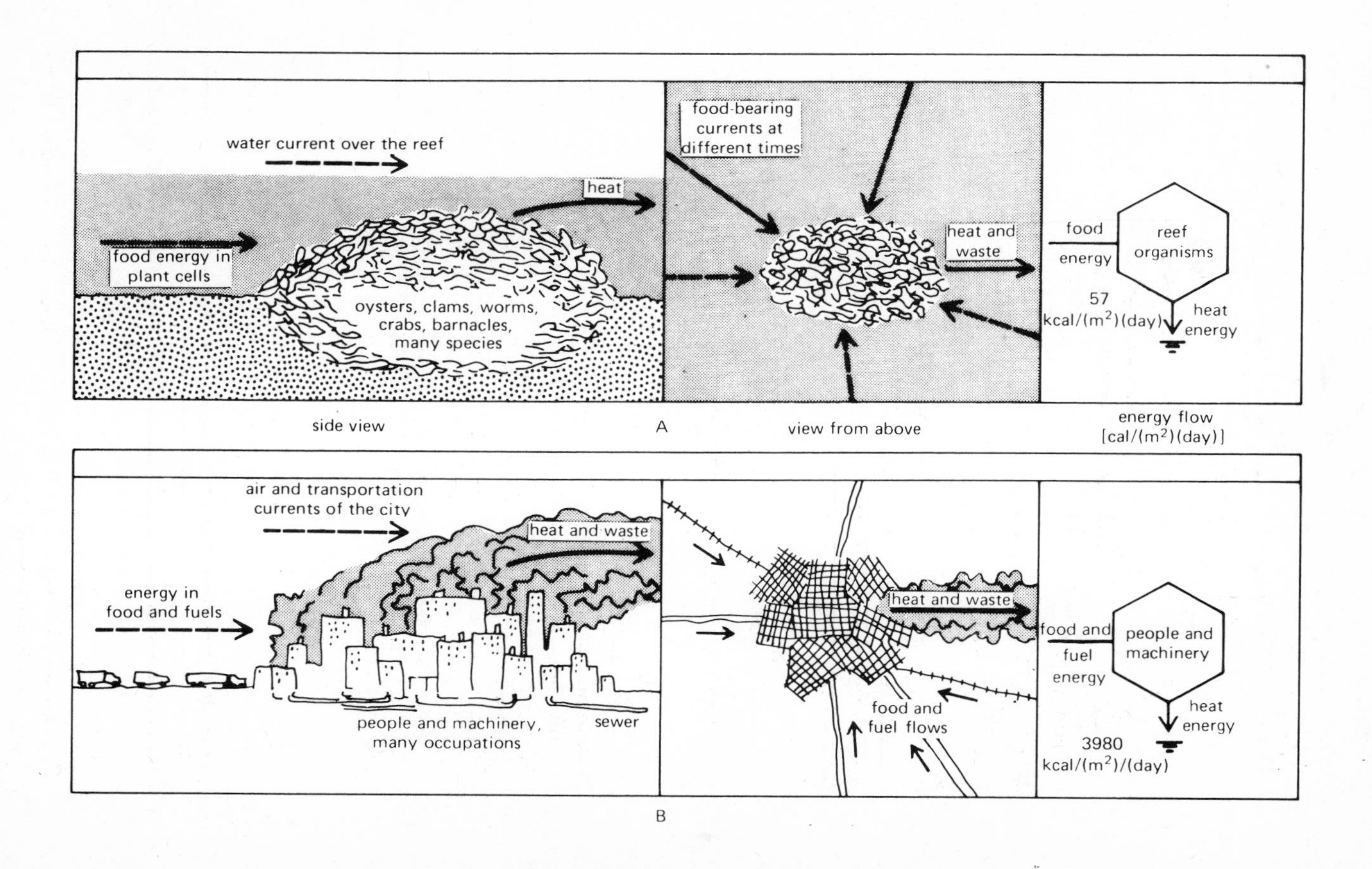

water current over the reef
heat
food energy in plant cells
oysters, clams, worms, crabs, barnacles, many species
food-bearing currents at different times
heat and waste
food energy
reef organisms
57 kcal/(m²)(day)
heat energy
side view
A
view from above
energy flow [cal/(m²)(day)]
air and transportation currents of the city
heat and waste
energy in food and fuels
people and machinery, many occupations
sewer
heat and waste
food and fuel flows
food and fuel energy
people and machinery
heat energy
3980 kcal/(m²)/(day)
B

manufacture food from simple inorganic substances and, secondly, a *heterotrophic* component (heterotrophic means "other nourishing"), which utilizes, rearranges, and decomposes the complex materials synthesized by the autotrophs. As shown in Figure 2-1, these functional components are arranged in overlapping layers with the greatest autotrophic metabolism occurring in the upper "green belt" where light energy is available, and the most intense heterotrophic activity taking place in the lower "brown belt" where organic matter accumulates in the soils and sediments. As already indicated, autotrophic and heterotrophic activity in the natural and seminatural landscape taken as a whole tends to be balanced in contrast to man's fuel-powered civilization. Even though many cities have extensive green belts (grass, trees), consumption of organic matter of high-energy value greatly exceeds production. It should be emphasized that there is nothing wrong or "bad" about cities being heterotrophic so long as they are linked with adequate autotrophic systems and can be supplied with necessary inflow of high-utility energy. As shown in Figure 2-2, nature does have her heterotrophic "cities," such as oyster reefs, but they are much lower-powered (compare 57 and 3980 kcal daily energy flow).

From another point of view it is convenient to recognize four constituents as comprising the ecosystem, as is also shown in Figure 2-1: (1) *abiotic substances and conditions of existence,* basic elements, compounds, and climatic regimes of the environment; (2) *producers,* the autotrophic organisms, largely the green plants; (3) the large *consumers* or *macroconsumers,* heterotrophic organisms, chiefly animals, that ingest other organisms or particulate organic matter; (4) the *decomposers* or *microconsumers,* heterotrophic organisms, chiefly the bacteria and fungi that break down the complex compounds of dead protoplasm, absorb some of the decomposition products, and release simple mineral nutrients usable by the producers as well as organic components which may provide food or which may be stimulatory (that is, vitamins) or inhibitory (that is, antibiotics) to other organisms.

You will note that this ecological classification of biotic components into three categories is based on modes of nutrition, that is, the principal source of energy utilized. In more technical language based on the root "troph" (= nourish) producers, consumers, and decomposers are often designated respectively as: autotrophs, phagotrophs (phago = to ingest), and saprotrophs (sapro = to decompose). Such an ecological classification is not to be confused with taxonomic classification (phyla, class, order, species, and so on), but there are parallels since the three modes of nutrition (photosynthesis, ingestion, absorption) are predominant in the taxonomic kingdoms of plants, animals, and fungi usually considered to be the three terminal branches of the evolutionary tree (see Whittaker, 1969). However, the ecological classification is one of function, not species as such. Many species occupy intermediate positions in the nutrition series, and still other species are able to shift their mode of nutrition. For example, some kinds of algae are able to function either as autotrophs or heterotrophs according to the availability of sunlight and organic matter.

The three functional types of living organisms comprise the biotic portion of an ecosystem. In an inventory sense the weight of organisms present at any one time is conveniently termed *biomass* (= living weight) or *standing crop.* As will be emphasized later, the size of the standing crop is not necessarily indicative of the level of activity; some ecosystems, such as a forest of large trees, have a large amount of relatively inert biomass.

It is also convenient to subdivide the nonliving or abiotic portion of an ecosystem into three components: (1) *inorganic substances,* the carbon, nitrogen, water, and so on that are involved in the material cycles of the ecosystem; (2) *organic substances*, the carbohydrates, proteins, lipids, humic substances, and so on that link abiotic and biotic; and (3) the *climate regime,* temperature and other physical factors that delimit the conditions of existence. Let us take a brief look at each of these essential ingredients.

Of the very large number of elements and simple inorganic compounds found at or near the surface of the earth, certain ones are essential for life. These are conveniently designated as *biogenic substances* or *nutrients.* Carbon, hydrogen, phosphorus, calcium, and potassium, among others, are required in relatively large amounts and hence are designated as *macronutrients*; these usually occur most abundantly in the form of simple compounds such as carbon dioxide, water, nitrates, and so on. Other elements, no less vital but required only in small amounts by living organisms, are known as *micronutrients.* There are at least ten of these that are essential to plants and most animals; these include a number of metal ions such as iron, manganese, magnesium, zinc, cobalt, and molybdenum. Still others are known or suspected to be essential for particular groups of organisms. While from 50 to 150 lb each of such macronutrients as nitrogen, phosphorus, and potassium are required to produce 100 bushels of corn (a better than average annual yield for an acre of good corn land), less than 0.1 lb of most micronutrients would be needed. Yet, a lack of micronutrient can hamper the productivity of an ecosystem just as much as would a shortage of a macronutrient. Without molybdenum, for example, microorganisms are unable to transform the nitrogen in the air into nitrates usable by plants.

The carbohydrates (sugars, starches, cellulose, and so on), the proteins (including amino acids, and so on), and the lipids (fats, and so on), which make up the bodies of living organisms also are dispersed widely in nonliving forms in the environment. These and hundreds of other complex compounds make up the organic component of the abiotic compartment. As the bodies of organisms decay they become dispersed into fragments of widely varying size, collectively called *organic detritus* (= product of disintegration, from the latin *deterere,* to wear away; the word detritus is also used in geology for the products of rock disintegration). Since the biomass of plants is usually greater than that of animals, and since plants usually decay more slowly than animals, detritus of plant origin is usually more prominant than that of animal origin. Organic detritus plays many important roles in the ecosystem, as will be noted in Chapter 3.

Organic matter in the environment occurs in a dissolved as well as a particulate form. As the breakdown of organic matter proceeds, materials called *humus* or *humic substances* are formed that are quite resistant to further decay, which means that they may remain for some time as a structural part of the ecosystem. Humus is the dark, yellow-brown, amorphous, or colloidal substance readily visible in soils, sediments, and suspended in the waters of streams and lakes (especially noticeable in swamp or bog water). Humic substances are difficult to characterize chemically. For those of you who have had a course in organic chemistry, we can say that they consist of chains of aromatic or phenolic benzene rings with side chains of nitrogen complexes and carbohydrate residues. The role that humic substances play in the ecosystem is not fully understood, but we do know that they contribute to soil properties favorable to plant growth. We also know that in large quantities they inhibit plant productivity. Under certain conditions, such as existed in past geological ages, organic matter of plant origin becomes completely fossilized to form coal, oil, and the other "fossil fuels" on which man's present fuel-powered societies now largely depend.

It is important to note that very few substances, inorganic or organic, are found exclusively in either the biotic or abiotic compartment of the ecosystem. Humic substances are not found inside living cells, but most everything else moves freely between organisms and environment. This is why we have to be careful what we "dump" into the environment, because whatever we put there could end up in our bodies. Unfortunately, the by-products of man's technology have recently tended to become increasingly poisonous, which means that we must take greater and greater care with waste disposal, or else redesign the manufacturing process to avoid such by-products, or both. Since removing poisonous chemicals from the waste disposal stream requires expensive energy, it would seem prudent to concentrate on the latter strategy, as advocated by Commoner in his widely read book *The Closing Circle* (1971).

We now come to the third category of the abiotic component of an ecosystem, the physical factors that determine conditions for existence for the organisms. On land, climate (temperature, rainfall, humidity, and so on) as well as chemical nature of the soil and underlying geological stratum, are major features that determine the kinds of organisms that are present, and indirectly, how well they are able to utilize available sun energy and energy subsidies. In aquatic ecosystems, temperature, salinity, and related chemical attributes of the body of water, and the nature of the sediments are major boundary conditions. Characteristic of the biosphere are a series of gradients of physical conditions, as for example: temperature gradients from arctic to tropics or from mountain top to valley; moisture gradients from wet to dry along major weather systems; depth gradients from shore to the bottom of bodies of water. Frequently, conditions and adapted organisms change gradually along the gradient, but often there are points of abrupt change or junction zones known as *ecotones* as, for example, prairie-forest junctions or intertidal zones on a seacoast. Once the nature of a gradient is understood we may often predict with considerable accuracy conditions and organisms present at a particular point in the gradient without actually having to make measurements or observations.

One of the remarkable properties of communities in which organisms have evolved together in groups (see page 167 for an explanation of coevolution) is their ability to compensate for changes in physical conditions (recall the principle of integrative levels, as discussed in Chapter 2). Thus, except under extreme conditions, different ecosystems are often able to maintain the same level of productivity under different conditions of temperature or other factors. For example, communities of kelp and other underwater seaweeds along the Nova Scotia coast are adapted to grow throughout the winter when nutrients and turbulent mixing are favorable even when water temperatures approach 0°C; as a result, annual productivity equals or exceeds that of tropical sea grass or reef communities (see Mann, 1973). In Chapter 7 we will consider in more detail the nature of factor compensation along gradients in the major ecosystems of the world.

THE MEADOW AND THE POND AS ECOSYSTEMS

The ecosystems illustrated in Figure 2-1 are contrasting types of sun-powered ecosystems, and thus emphasize basic similarities and differences. A terrestrial ecosystem (illustrated by the field shown on the left) and an open-water aquatic system (illustrated by a lake or the sea as shown on the right) are populated by entirely different kinds of organisms, with the possible exceptions of a few kinds of bacteria that may be able to live permanently in either situation. Yet the same basic ecological components are present and function in much the same manner in both types of ecosystem. On land, the autotrophs are usually rooted plants ranging in size from grasses and other herbs that occupy dry or recently denuded lands to very large forest trees adapted to moist lands. In deep water systems the autotrophs are microscopic suspended plants called *phytoplankton* (phyto = plant; plankton = floating), which belong to several different classes of algae. They include: (1) the diatoms, tiny plants with silicon shells; (2) green flagellates that move about propelled by rapidly beating shiplike flagella; (3) the green algae, which may occur as single cells, colonies, or filaments of cells; and (4) the blue-green algae, some of which have gelatinous capsules and thrive on organic pollution, thus clogging public water supplies and creating nuisances in recreational lakes. As would be expected, shallow water ecosystems are occupied by mixtures of macroscopic plants and microscopic algae.

Because of size differences in plants, the biomass, or standing crop, of terrestrial and aquatic ecosystems may be widely different. Plant biomass in terms of grams of dry matter per square meter may be 10,000 or more in a forest in contrast to less than 5 in a pond, lake, or ocean. Despite the size discrepancy, 5 g of tiny plants are capable of manufacturing as much food in a given period of time as are 10,000 g of large plants given the same quantity of light, minerals, and energy subsidies. This is because the rate of metabolism of small organisms is very much greater per unit of weight than that of large organisms. Furthermore, large land plants are mostly composed of woody tissues that are relatively inactive; only the leaves are active in photosynthesis, and in a forest leaves comprise only about 1 to 5 percent of the total plant biomass.

This is a good place to introduce the concept of *turnover* as a first step in relating structure to function in an ecosystem. We can think of turnover as the ratio of the standing state (that is, amount present) of biotic or abiotic components to the rate of replacement of the standing state. For example, if the biomass of a forest is 20,000 grams per square meter (g/m^2) and the annual growth increment is 1000 g, then the ratio 20/1 can be expressed as a *turnover time* or *replacement time* of 20 years. The reciprocal, that is, $1/20 = 0.05$, is the *turnover rate*. In a pond the turnover time for phytoplankton would be measured in days rather than years.

It is particularly important to have information on turnover rates between biotic and abiotic compartments when it comes to evaluating the impact of mineral nutrients or other chemical components in an ecosystem. It is more important to know how fast materials are moving along the pathways between organisms and environment than it is to know the total amount present. Thus, a soil might contain a large amount of phosphorus, but if it is not available to organisms, perhaps because it is in an insoluble form, then it might as well not be there. We have already made note of man's tendency to extract materials from the environment and return them to the environment in poisonous forms; man also, often inadvertently, returns them in unusable form. Then it is like the man marooned in the middle of the sea—there is "water, water everywhere, but not a drop to drink."

The Pond as an Ecosystem After a number of years of experimentation in the teaching of a beginner's course in ecology we have found that a series of field trips to a small pond provide a good beginning for the "lab" part of the course. A pond has a distinct boundary and is thus a recognizable unit in terms of both structure and function, even though it is not a closed system. Just as the pond frog is a classical type for the introductory study of the animal organism, so the pond itself proves to be an excellent type for the beginning study of ecosystems. A small pond "managed" for sportfishing is the best type to start with because the number of species of organisms present is small, but almost any pond, even a marine embayment will do. The four basic components can be sampled and studied without the beginner becoming lost in too much detail. Furthermore, measurement of oxygen changes over a diurnal cycle provides a ready means of measuring the rate of metabolism and of demonstrating the interaction of autotrophic and heterotrophic components in the ecosystem as a whole.

The "dissecting tools" that the ecologist uses in his study of ponds are shown in Figure 2-3A, and some of the laboratory apparatus needed for quantitative measurements are shown in Figure 2-3B. In a class study, students may be grouped into teams, each of which is assigned to the job of sampling a major component or making a key measurement. One team, for example, dissects out the producers by taking a series of water samples with a special sampler that traps a column of water at any desired depth (Figure 2-3A). Back in the laboratory, part of the water samples are filtered to concentrate the tiny phytoplankton organisms for microscopic study and counting. Another part of the samples is then passed through a very tight filter that removes all of the organisms (Figure 2-3A); the dried filter with the organisms is then placed in acetone to extract the chlorophyll and

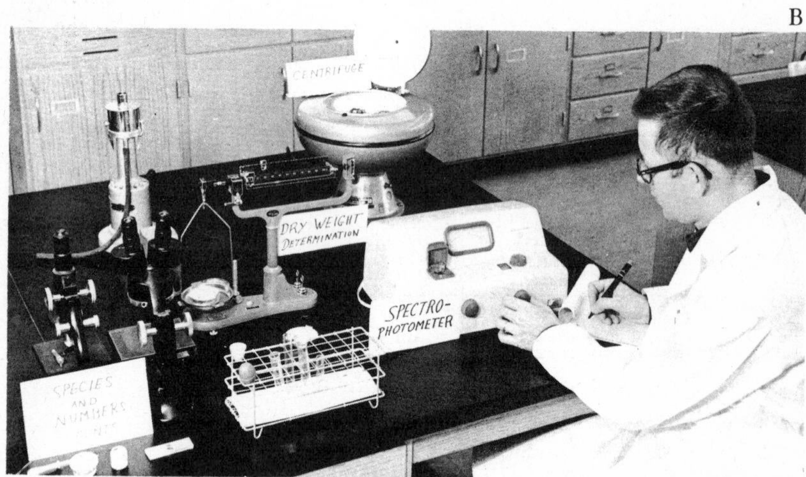

Fig. 2-3 The field and laboratory tools used by an ecology class in the study of pond ecosystem. (A) Field equipment including sampling devices for water, chlorophyll, plankton, bottom fauna, and fish together with apparatus for measuring oxygen metabolism of the pond. (B) Laboratory apparatus for further study of samples collected in the field.

other pigments. The resulting clear green solution can then be placed in a photoelectric spectrometer (Figure 2-3B) to determine quantitatively the actual amount of chlorophyll and other pigments. The total quantity of chlorophyl in a water column, or in a community in general on an area basis (that is, per square meter), tends to increase or decrease according to the amount of photosynthesis. Therefore, chlorophyll per square meter (m^2) is an index of the food-making potential at a given time, since it adjusts to light, temperature, and available nutrients. A general model for chlorophyll in ecosystems will be presented later in this chapter. Chlorophyll data can also be used to estimate the living weight or biomass of producers, while the amounts of other pigments tell other stories should we wish to go more deeply into the study.

Similarly, other teams obtain numbers, kinds, and weights for the consumer groups. *Zooplankton,* which are the small consumers associated with the water column, can be sampled by dragging a plankton net made of very fine-mesh silk through the water, fish can be sampled by seining, and small animals living on and in the bottom sediments can be quantitatively collected with a "grab" built on the principle of a steam shovel (Figure 2-3A). From these data a picture of the structure of heterotrophic populations is obtained.

As already emphasized, simply inventorying the components of an ecosystem does not tell us much about what goes on in the system; for full understanding it is also necessary to make measurements of the rate of energy flow, the rate of nutrient exchange, and other functional properties. For example, oxygen changes in the water column can be measured as a means of assaying the metabolism. One way to do this is to suspend light and dark bottles in the pond to measure oxygen changes resulting from autotrophic and heterotrophic metabolism, respectively, as shown in Figure 2-4. A portion of a sample of water from each of several levels is placed in glass bottles. One or more bottles are covered with aluminum foil or black tape so that no light can reach the sample; these are called the *dark* bottles, in contrast with the *light* bottles that have no such cover. Other bottles are "fixed" with reagents immediately so that the amount of oxygen in the samples at the beginning of the experiment can be known. Then pairs of light and dark bottles are suspended in the pond at the levels from which the water samples were drawn. At the end of the 24-hour period the string of bottles is removed from the pond and the oxygen in each is "fixed" by addition of a succession of the three reagents: manganous sulfate, alkaline iodide, and sulfuric acid. This treatment releases elemental iodine in proportion to the oxygen content. The water in the bottles is thus now brown in color; the darker the color the more oxygen. The brown water is then titrated in the laboratory

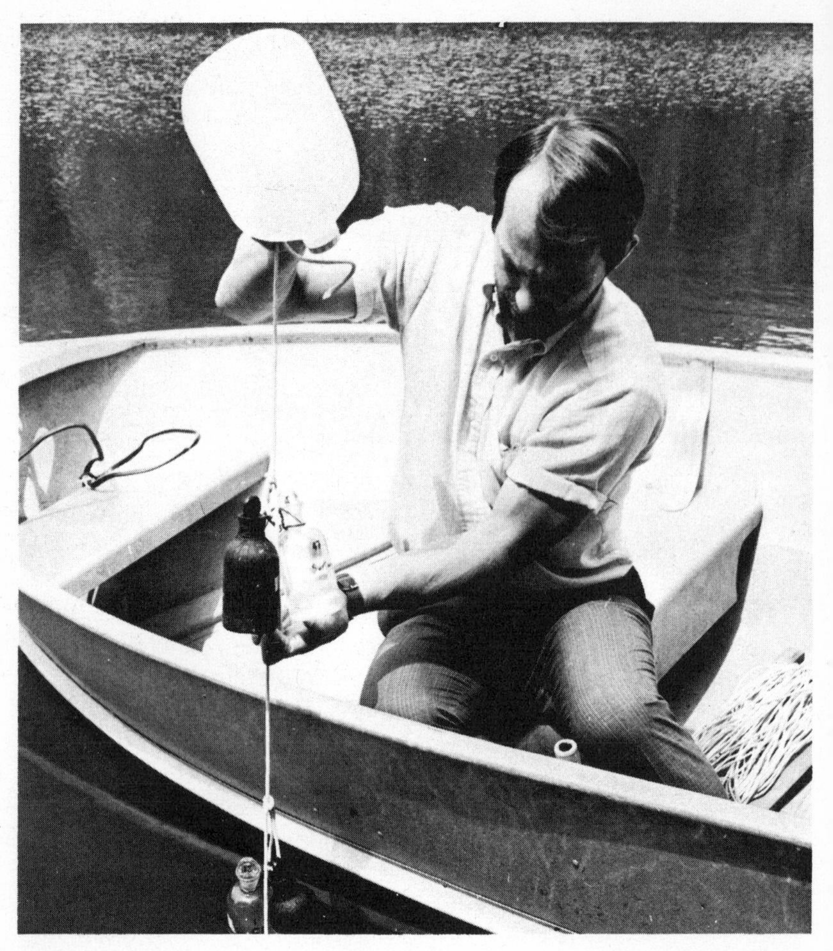

Fig. 2-4 Measuring the metabolism of a pond by the light-and-dark bottle method. Pairs of black and transparent bottles are lowered into position, while the plastic jug acts as a float to hold the pairs at the desired levels for measurement of oxygen changes. See text for explanation.

by adding sodium thiosulfate (the "hypo" used to fix photographs) until the color disappears. The volume of sodium thiosulfate needed can be calibrated to indicate the concentration of oxygen in milligrams or milliliters per liter; milligrams per liter is also parts per million, another way in which oxygen content of water is expressed. This chemical method of measuring oxygen in water is known as the *Winkler method*; it has been and continues to be the standard method, although newer electronic methods involving the use of oxygen electrodes offer advantages, especially where we wish to have a continuous record of change over time.

The decrease of oxygen in the dark bottles indicates the amount of respiration (that is, heterotrophic metabolism) in the water column whereas the oxygen change in the light bottles indicates the net photosynthesis (that is, net result of photosynthesis and respiration); the two quantities added give an estimate of total photosynthesis or total food production for the 24-hour period, since oxygen production by green plants is directly proportional to fixation of light energy. One method of calculating the photosynthetic rate of the water column on a square meter basis is to average values for each meter level and convert to oxygen per cubic meter (a simple shift of the decimal since milligrams per liter equal grams per cubic meter); the values for each meter level when added give an estimate of total oxygen production per square meter of pond surface. In the simplest case, if bottles had been placed at 0.5, 1.5, and 2.5 m deep, then each pair could be considered as sampling the first, second, and third cubic meter; the sum of these would give an estimate for a column 3 m deep. Alternatively, a graph of bottle values plotted against depth can be constructed and the area under the curve used to estimate the column.

A factor of 3.5 can be used as an approximate conversion of grams of oxygen produced to kilocalories of organic matter fixed by the plants in photosynthesis. Thus, if the accumulated change in oxygen in a square meter of water column was +3 g in the light and −2 g in the dark, the total, or gross, production would be 5 g, or 16.5 kcal/m^2/day; of this amount 6 were used (that is, respired) by the plankton community, leaving 10.5 to be used or stored in the bottom of the pond. A cloudy day or organic pollution could result in more energy used than produced (that is, more oxygen consumed in the dark bottle than produced in the light bottle). Thus, measuring the oxygen metabolism provides not only an index of the energy flow, or power level, as discussed at the beginning of this chapter, but also an indication of balance between autotrophic and heterotrophic activity. As will be discussed in Chapter 6, most small ponds are not metabolically balanced over the annual cycle, but are changing in structure and function with time.

Where phytoplankton density is very low, as in large deep lakes or the open ocean, the sensitivity of the light and dark bottle method can be greatly increased by adding a radioactive carbon tracer to the water in the bottles. After an interval of time the phytoplankton is removed by a filter that is placed in a detector to determine the amount of radioactive carbon fixed (that is, transferred from water to phytoplankton). This method, which indicates the net photosynthesis, is widely used in oceanographic work. At sea it is not necessary to resuspend bottles in the water and stand by for 24 hours; the samples can be subjected to the light and temperature conditions of the sea on the deck of the ship as it moves to a new sampling location.

In another approach the whole pond can be considered as a dark and light bottle. If oxygen measurements are made at 2- or 3-hour intervals throughout a 24-hour cycle, a diurnal curve may be plotted that shows the rise of oxygen during the day when photosynthesis is occurring and the decline during the night when only respiration is occurring. The daytime period is equivalent to the light bottle and the night to the dark bottle. The advantage of this diurnal curve method is that photosynthesis of the whole pond including plants growing on the bottom (which would not be included in bottles) would be estimated. The difficulty is that physical exchange of oxygen between air and water and between water and sediments must be estimated to obtain the correct estimate for oxygen production of plants in the pond. Usually, the bottle methods give a sort of minimum and the diurnal curve a sort of maximum estimate.

In addition to community structure and community metabolism, water chemistry is a third ecosystem property that should receive some attention in a class study. Recent advances in colorimetry and spectrophotometry have made water analysis relatively easy, assuming one has funds to purchase a colorimeter or spectrophotometer and ready-made reagents, or if one has access to a water chemistry laboratory. Even without such resources some idea of physical conditions of existence can be obtained with inexpensive water test kits of the type homeowners use to check out the water quality of swimming pools. In any event, the principle is the same; for a given substance to be inventoried, reagents are added to water samples to produce a color, the intensity of which is proportional to the concentration of substance in question. Nitrate nitrogen, ammonia nitrogen, and phosphate phosphorus are macronutrients well worth measuring. Temperature, pH, transparency (turbidity), and total alkalinity (hardness) are easy to measure and provide key information on the chemical state of nutrient elements and their availability to organisms. The role that chemical and physical factors of the environment play in limiting or enhancing the metabolism and diversity of the ecocystem will be discussed in Chapter 5.

Concentrations of some of the metallic ions, such as iron, copper, zinc, lead, and chromium in the water, in bottom sediments and/or in bottom organisms and fish are of especial interest as indicators of industrial pollution. Measurements of these elements in an unpolluted pond where concentrations should be very low provides a good yardstick for assaying the condition of streams or ponds suspected of being polluted by the waste products of industrial and commercial operations. Comparison of number and variety of organisms in the two situations provides a means of assessing the impact on the pollutants in the biotic community; more about this later. Thus, a visit to polluted waters (which should not be hard to find these days) is a natural follow-up to the pond study. The contrast can be educational, to say the least!

A meadow or old field is also a good place to start the study of ecology. Different sampling procedures, of course, would be used to inventory the biotic community, some examples of which are shown in Figure 2-5. The total, or gross, productivity is more difficult to measure in the terrestrial environment because of the thermal problems created by enclosing vegetation, and the greater difficulties of measuring gaseous exchange in air media (carbon dioxide rather than oxygen exchange would be assayed). However, an estimate of net production can be made in herbaceous communities by harvesting, weighing, and summing the living plant material produced and the dead material (litter or detritus) accumulated during the growing season. As already indicated, terrestrial plants, in contrast to phytoplankton, store or accumulate energy over longer time periods to a degree related to the turnover time of the producer biomass, and the timing and amount of consumption by animals. If a meadow or field is not being grazed by large animals, most of the organic matter produced during the growing season is still present at the end of the season.

TRACERS AS AIDS IN ASSESSING FUNCTION

Just as the microscope extends our power of observation of the details of structure of components in the ecosystem, so tracers extend our power of observation of function. By tracers we mean small amounts of easily detected substances that can be used to follow and quantify the flow of materials or movement of organisms not otherwise visible or detectable by ordinary means. Tracers can take many forms ranging from dyes used to trace water movement to isotopes that can be used to measure nutrient exchange between organisms and environment because they are easily detectable by special instruments. Justifiable concern about radioactive pollution has overshadowed the fact that radioactive tracers provide valuable tools for study. Many vital elements have radioactive isotopes with short half-lives (that is, decay to nonradioactive form in a few days), which can be detected and measured in such small quantities so that tracer amounts introduced into the system will have no meas-

A

B

Fig. 2-5 Sampling methods in a study of a terrestrial ecosystem. (A) Estimating net productivity and species diversity of a one-year-old field (cultivated field abandoned for one growing season) by harvesting square meter samples of the autotrophic standing corp. Material in paper bags will be taken to the laboratory for sorting into species and weighing. (B) Estimating the short-term photosynthetic rate by measuring the uptake of CO_2 by plants within the chamber. (Photographs courtesy of the Institute of Ecology, University of Georgia.)

urable effect on the process being measured or the organisms present. And an isotope does not have to be radioactive to be useful as a tracer. The nonradioactive isotope nitrogen-15, for example, has been very useful in the study of the all important nitrogen cycle.

Experiments with radioactive tracers provide excellent laboratory experiments to follow a field study of the pond. For example, paired bottles of filtered pond water can be set up, each "spiked" with tracer amounts of radioactive phosphorus (^{32}P).[1] In one of a pair of bottles a gram or two of a large, leafy submerged aquatic plant is placed, and in the other, a known weight of filamentous or phytoplankton algae. The uptake by the plants is easily monitored by withdrawing and filtering small samples of water at intervals over a period of several hours, and counting the samples in a suitable detector; decrease in radioactivity of the water provides a relative measure of the amount of phosphorus moving into the plant biomass. The much more rapid uptake per unit weight of small plants, as compared with the large one, provides a dramatic illustration of the differences in nutrient turnover rates, as previously discussed.

THE TROPICAL REEF: A COMPLEX ECOSYSTEM We would do well to close the general discussion of the study of natural ecosystems with an example that illustrates the value of studying the whole ecosystem as well as the component parts, even when the system is much more complex than a small fish pond or a field. A tropical biotic reef (Figure 2-6) represents one of the most beautiful and well-adapted ecosystems to be found in the world. Corals, small animals with hard calcereous skeletons, and calcareous algae build up the reef substrate which is the home of numerous organisms. As shown in Figure 2-6, the animals are closely associated with plants. Embedded in the tissues of the coral, and also in and on the skeleton of many animals and the general calcareous substrate are numerous algae. If supplied with abundant zooplankton food, some coral species can be maintained in laboratory tanks without the algal associates. However, when the metabolism of a whole reef is measured (as, for example, by measuring diurnal changes in oxygen as water passes over the reef—a modification of the method just described for assaying the metabolism of a pond), the input-output budget indicates not enough animal food suspended in the water to completely support the corals. In such a situation there must be supplemental sources of food, perhaps that produced by algal associates. Tracer experiments have shown that exchanges of organic matter between plant and

[1] Small amounts of phosphorus-32 and other radioactive nuclides can be purchased for educational use without a license. A license is required for purchase of large amounts of radioactive materials for laboratory use or for any use in the field.

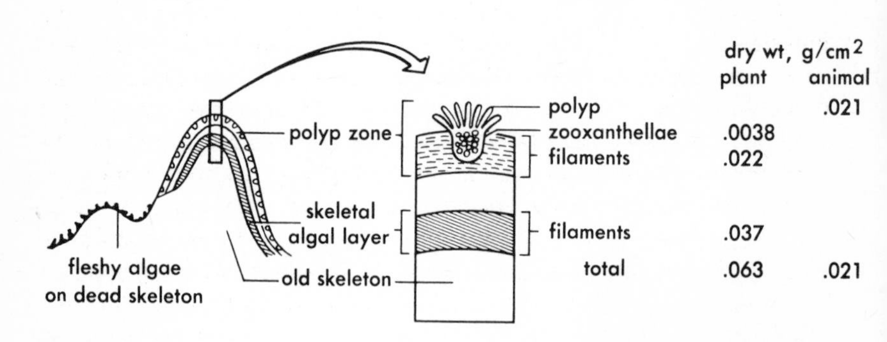

Fig. 2-6 A coral reef, one of the most complex and productive of natural ecosystems. Upper photo is a general underwater view showing the irregular masses, "heads," and branched, treelike structures produced by corals of different species. What we see at a distance are the skeletons (often brightly pigmented), embedded in which are thousands of living animals and plants that have created the reef. The inset is a closeup taken at night when tentacles of the individual coral animals or "polyps" are expanded. The diagram is a cross section of a coral head or colony showing the intimate association between the coral animal and several types of algae. (Photographs courtesy of Dr. Carlton Ray, New York Aquarium. Diagram redrawn from H. T. and E. P. Odum, *Ecological Monographs*, Vol. 25, 1955.)

animal tissues within the colony do occur. Also, it has been clearly demonstrated that mineral nutrients are recycled back and forth between animal and plant components so the colony does not require a high rate of fertilization from without. These discoveries indicate that, in nature, coral animals and algae are metabolically linked and dependent on one another. The history of recent research on coral reefs bears out the point we have already emphasized: The behavior of an isolated component (coral in a tank) may not be the same as the behavior of the same component in its intact ecosystem (the reef) where available energy sources and nutrient constraints may be quite different. And the corollary to this: To understand the ecosystem, the whole as well as the part must be studied.

THE CITY AS AN ECOSYSTEM Now let us consider the ecology of the fuel-powered city within the same frame of reference as taken in our discussions of ponds, meadows, forests, and reefs. Referring back to Figure 2-2, we note again that the city as the oyster reef, but not as the coral reef, is a heterotrophic ecosystem dependent on large inflows of energy from outside sources. Actually, most cities contain large numbers of trees, substantial areas of grass and shrubs, and in many cases, lakes and ponds—so they do have an autotrophic component or green belt. However, the organic production (converted solar energy) of the city green belt does not contribute appreciably to the support of people and machines that so densely populate the urban-industrial area. The urban forests and grasslands do have an enormous aesthetic value and they do contribute indirectly to pollution abatement by reducing noise, carbon dioxide, and other waste products of fuel consumption. But fuel and labor expended in watering, fertilizing, pruning, removing wood and leaves, and other work required to maintain the city's private and public green belts, adds to the energy (and money) cost of living in the city. It was already noted in our discussion of the basic kinds of ecosystems that sunlight is more of a liability than a benefit in the twentieth century city, but this situation may change when fuel becomes limited in supply or high in price.

In Figure 2-7, an American city is compared with a natural ecosystem of comparable size, namely a large lake. This figure is a combination pictorial and tabular model in three dimensions: A. structure (zonation or "land use"); B. populations of organisms; and C. major inputs and outputs of energy and materials. The hypothetical city has a population of one million people, a population density of 11.2 per

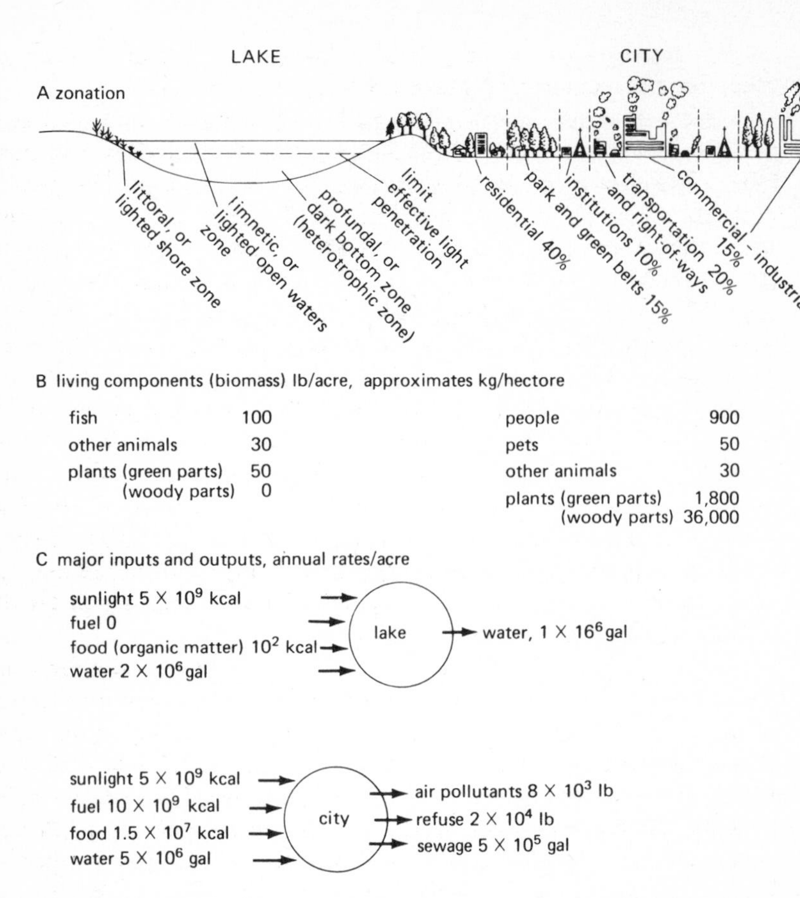

Fig. 2-7 Comparison of a natural autotrophic ecosystem (lake) and a fuel-powered, heterotrophic city in regard to: (A) zonation (structure); (B) population density (biomass of living components); (C) Major inputs and outputs of energy, water, and wastes.

acre, and a land-use pattern as shown in the upper left diagram. The latter two attributes represent the average statistics for seven American cities (New York, Chicago, Philadelphia, Los Angeles, Detroit, Cleveland, and Pittsburgh), as tabulated by Abrams (1965). At a density of 11 per acre a city of a million would occupy about 90,000 acres (about 140 square miles or 36,000 hectares). The hypothetical 90,000-acre lake is modelled to resemble a large, shallow natural lake of moderate fertility. The data in sections B and C of the model are expressed in terms of quantities per acre and per hectare (in parentheses), since total amounts for the large city and lake come to very large figures that would be more or less incomprehensible to the reader. However, to calculate totals all you have to do is multiply everything in sections B and C by 9×10^4.

Now let us compare the general layout of the city and lake. Both ecosystems tend to exhibit a concentric pattern of zones which, however, often overlap or interdigitate in a complex manner. The peripheral shallow water or shoreward zone of the lake, where light penetrates to the bottom, is known as the *littoral zone*; rooted water plants such as pond weeds, cattails, water lilies, and so on, are often found here. The outer zone of the classic city is generally residential with rooted plants also conspicuous (but, as indicated above, these are there more for show than for utility). The inner open water zone to the depth of light penetration effective for photosynthesis, is known as the *limnetic zone*. The littoral, together with the limnetic zone, make up the autotrophic zone of the lake. It is here that solar energy is converted to organic "fuels" that support the inhabitants of the lake (including any fish harvested by man or other animals). The rest of the lake, including the deeper waters and the large area of bottom that is beyond light penetration, is the *profundal zone*, and comprises the strictly heterotrophic part of the ecosystem.

The city generally has a distinct core of energy-consuming commercial development. However, the highest energy consumption occurs in the industrial areas which may be located in the city center, or in islands or bands in or around the city. A characteristic feature of cities is the network of transportation arteries which, together with the right-of-ways, take up a surprisingly large portion of the land area (20 percent, as shown in Figure 2-7). Transportation of materials and people leads to intense and rather inefficient energy consumption in terms of incomplete combustion of fuel that leaves poisonous by-products in the air. Transportation energy use is a major cause of the air pollution that plagues all large cities in affluent countries. In contrast, transportation of materials in the lake is accomplished by waves and currents powered by the wind and the sun. Those organisms, such as fish, which do travel extensively, do so under their own power. Thus, there is little pollution resulting from circulation and transportation in the lake. A point to emphasize is that both population density and energy consumption density are very unevenly distributed in both the solar-powered lake and the fuel-powered city. Nevertheless, average figures do provide a useful overall comparison of the two.

Surprisingly enough the amount of life on a per-unit-area basis is not greatly different in the city and in the lake, as shown by the comparison of Figure 2-7B. Fish outnumber people, but people outweigh fish by about 10 to 1. When we consider the large number of pets, rats, birds, insects, and so on that inhabit the city the biomass of animals is greater than in the lake; and there is a greater biomass of plants in the city even if we consider only the photosynthetically active green parts. Although 25 percent or so of the land area of a city may be completely devoid of plants the total cover of trees, shrubs, and grass approximates that of a natural terrestrial ecosystem or a rural area covered with a mixture of grassland and forest. In satellite photos of the earth's surface only the commercial and industrial areas of cities stand out; much of the sprawling urban area resembles, at first glance at least, the surrounding countryside. As already emphasized, it is the level and type of energy and material flow that makes the big difference between the fuel-powered city and the solar-powered natural ecosystem, and this can best be appreciated at ground level. Comparison of major inputs and outputs, as shown in Figures 2-7C, will serve to bring out these differences.

The huge input of fuel energy required to support the dense population of machines, and to heat and cool the buildings and homes, has no counterpart in the lake system. The small amount of organic matter imported into the lake from the watershed might be considered in this category, but we have placed this in the food compartment in our graphic model. Remember that we are considering a lake with a natural watershed; urbanization of the watershed can alter the energy and material budgets of the lake quite considerably, making it much more heterotrophic. In Table 2-2 energy consumption density is estimated for large cities, industrial regions, whole countries, and the world as a whole. For our hypothetical model city (Figure 2-7C) we have set the annual consumption level at 10 billion kcal per acre (about $2.5 \times 10^6/m^2$), which is about halfway between a concentrated city such as New York and a more spread-out city such as Los Angeles. The per capita consumption rate for such a hypothetical city would be about 8.9×10^8 ($10^{10} \div 11.2$) or something more than 10 times the national per capita average of 86×10^6, as cited on page 120. Consumption per acre in the city, on the other hand, is more than 1000 times that of the United States as a whole, which is about $1.8 \times 10^3/m^2$ or 7.3×10^6/acre (see Table 2-2).

The reason for citing these large numbers is to emphasize that in terms of energy metabolism cities are pinpoint "hot spots" in the biosphere's surface. It is particularly important to note that although energy consumption in the largest cities exceeds the local solar energy input, mankind's burning of fuels is yet but a drop in the bucket compared to solar input on a global basis (see footnote, Table 2-2). But remember that solar is low utility (that is, in terms of work capacity) and fuel is high utility energy. Possible effects of the increasing intense fuel consumption on local and global climates and heat balances will

Table 2-2. Energy Consumption Density Directly Related to Man's Use of Fuels

		(kcal m^{-2} $year^{-1}$)[a]
a	CITIES	
	Manhattan (New York City Center)	4.8×10^6
	Tokyo	3.0×10^6
	Moscow	1.0×10^6
	West Berlin	1.6×10^5
	Los Angeles	1.6×10^5
b	LARGE INDUSTRIALIZED REGIONS	
	German Industrial Region	7.7×10^4
	Los Angeles Basin	5.7×10^4
	Japan (whole country)	2.3×10^4
	United Kingdom	9.2×10^3
	14 Eastern States, U.S.A.	8.4×10^3
	United States (whole country)	1.8×10^3
c	WORLD AVERAGE	100

[a] Compare these figures with the solar energy that reaches the earth's surface, which is somewhere between 1 and 2×10^6 kcal m^{-2} $year^{-1}$, depending on latitude.

be discussed in the next chapter, as will the uncertain prospects for running cities on solar energy.

In addition to fuel, the city must import all its food, in contrast to the lake where most, if not all, food required to support the organisms is grown within the ecosystem. The estimated import of food, as shown in Figure 2-7C, is greater than that required to support the 11.2 people and their pets that live on the urban acre, because much food is wasted. What the rats and other scavengers do not get goes in the garbage that forms part of the solid waste output, an estimate of which is shown in Figure 2-7C. Many cities in Europe are trying to convert this waste into fuel, soil mulch, or other energy-saving uses, and we can expect a similar effort in America soon.

The output of many acres of agricultural land is required to supply the city acre with food. As of 1973 about two acres are required to supply the rich diet of an American citizen, which means that some 22 acres of agro-ecosystems are required to feed the people inhabiting an acre of our hypothetical city. As already noted (see especially Table 2-1), the agro-ecosystem is heavily "fuel-subsidized" so large amounts of fuel are consumed outside the city to produce the food, an energy requirement that does not show up directly in the city energy budget. Recent estimates of the calories of fuel required to produce a calorie of food produced by different crop systems is shown in Figure 8-2. Although the use of land and fuel is much less lavish in most other parts of the world, many densely populated countries, such as Japan, do not have enough food-producing areas to support the population, so food must be imported from other countries. This brings up a point that we will come back to: Densely populated, high energy-consuming areas require low-density, energy-producing areas of much larger extent to support them. Thus, whether an area is judged to be over-populated or not depends not only on the socioeconomic consequence of crowding (that is, on population density, per se), but also on the capacity and availability of energy sources which may be located in distant regions.

The city's prodigious consumption of energy is coupled with large inflows of water and other materials, and large outflows of polluted water, solid wastes, air pollutants, and heat. Energy consumption and material flows are inseparably linked; the larger the flow of energy into the city the larger the inflow of materials and outflow of wastes, but the relationship is not linear since utilization may be improved at intermediate volumes. Water, metals, and so on are absolutely essential to the conversion of fuel energy to useful goods and services. Without water, for example, a city would quickly choke to death no matter how much oil or other concentrated energy is available. Likewise, wastes of degraded energy and end products of material fabrication are a thermo-dynamic inevitability of energy conversion, as will be documented in the next chapter. Unfortunately, as long as energy and resources are abundant and available at low cost there is little economic incentive for conservation, so resource use tends to be wasteful, producing an additional increment of wastes above that which is inevitable. Therefore, the output quantities, as estimated for the current American city (Figure 2-7C) could be reduced, and is reduced in other geographical areas where fuel is scarce. But there is a cost. For example, to clean up and recycle waste water, substantial amounts of fuel energy and tax dollars would have to be diverted from other uses to do this work. It is cheaper to let nature's hydrologic and photosynthetic systems (both of which run on solar energy) do most of this work free, but this is feasible only if there are no other large cities upstream and downstream. Capacity for waste treatment can be increased by judicious disposal of wastes on land as well as in water areas. Accordingly, the ecologic and economic budgets of a city are determined not only by energy consumption density and resource availability, but also by geographical location. A city located in a large matrix of seminatural environment is one thing; cities crowded back-to-back results in quite a different situation. We have already called attention to the fact that large cities are mostly located on free natural sewers.

The lake resembles the city in having a large "throughput" (that is, input-output) of water, but water quality is not so rapidly degraded by the lake in sharp contrast to what happens to water as it passes through the city. There are evaporative losses of water from natural ecosystems which match the "consumption" of water (that is, water lost in transit) in the city, resulting in a lower output than input for both types of ecosystems. Although rainfall is a useful "subsidy" to the lake it becomes a costly nuisance in the city. As is the case with sunlight, rain is not only little used by the city but it causes expensive trouble in terms of storm sewer maintenance and flood damage. Only in the driest climates is runoff from roofs or other artificial catchbasins used for drinking purposes. It usually is more convenient and cheaper for the city to get its water from a natural watershed.

To summarize, both the city and the lake require large watersheds, but, in addition, the city requires a large "foodshed" and "fuelshed," that is to say, distant areas that supply energy. The far greater rate of energy conversion in the city results in a waste output that stresses any kind of system (whether natural or man-made) that is located downstream or downwind. It is, as stated at the beginning of this chapter, the fuel-powered ecosystem can function only as a consumer (heterotroph) in the matrix of the solar-powered biosphere. Our challenge is to see that the city not only does not become a malignant parasite, but does become a more benevolent symbiont with its surroundings. We hope the data and examples presented in this chapter have convinced you that the fuel-powered city and solar-powered countryside are best viewed as, and in the future managed as, coupled systems. Unfortunately, present-day political and economic procedures are set up to deal with these two systems as if they were separate entities. Political conflicts between urban and rural seem to be an inherent pattern of human behavior even though there is no logic to it.

XONOMIC COMPONENTS IN THE ECOSYSTEM; HE ECOLOGICAL NICHE

We are all aware that the kinds of organisms to be found in both rural and urban areas in a particular part of the world depend not only on the local conditions of existence—that is, hot or cold, wet or dry—but also on geography. Each major land mass as well as the major oceans have their own special fauna and flora. Thus, we expect to see kangaroos in Australia but not elsewhere; or hummingbirds and cacti in the New World but not in the Old World. And the different continents are the original home of different races of human beings and different kinds of domesticated plants and animals. The fascinating story of adaptive radiation is considered in more detail in other volumes of the *Modern Biology Series* that deal with animal and plant diversity. From the standpoint of the overall structure and function of ecosystems, it is important only that we realize that the biological units available for incorporation into systems vary with the geographical region. The word *taxa* is a good term to use in this connection when we wish to speak of orders, families, genera, and species without wishing to designate a particular taxonomic category. Thus, we can say that both local environment and geography play a part in determining the taxa of an ecosystem. As already indicated, the type and level of energy plays an important role in determining the kinds as well as the numbers of organisms present. As will be discussed later, the biotic community itself may play an important role in this regard.

What is not always so well understood is that ecologically similar, or ecologically equivalent, species have evolved in different parts of the globe where the physical environment is similar. The species of grasses in the temperate, semiarid part of Australia are largely different from those of a similar climatic region of North America, but they perform the same basic function as producers in the ecosystem. Likewise, the grazing kangaroos of the Australian grasslands are ecological equivalents of the grazing bison (or the cattle that have replaced them) on North American grasslands since they have a similar functional position in the ecosystem in a similar habitat. Ecologists use the term *habitat* to mean the place where an organism lives, and the term *ecological niche* to mean the role that the organism plays in the ecosystem; the habitat is the "address" so to speak, and the niche is the "profession." Thus, we can say that the kangaroo, bison, and cow, although not closely related taxonomically, occupy the same niche when present in grassland ecosystems.

In recent years professional ecologists have become intensely interested in quantifying the concept of the ecological niche in terms of a set of conditions within which each kind of organism can operate (the fundamental niche) or does operate (the realized niche). In this manner "niche width" and "niche overlap" between two or more kinds of organisms can be compared. The reason for such interest stems from the discovery that the way in which taxa divide up available space, energy, and resources has a profound influence on the evolution of structure and behavior and on the origin and extinction of species. We will touch but briefly on these matters in this book, but if you wish to read more we suggest you start with the review by Whittaker, Levin, and Root (1973) and the book entitled *Geographical Ecology* by MacArthur (1972).

Man, of course, has a considerable influence on the taxonomic composition of many ecosystems, not only urban ones but remote ones in which he may be but a minor inhabitant. We might think of his efforts to remove or introduce species as a sort of ecosystem surgery;

sometimes the surgery is planned, but too often it is accidental or inadvertent. Where the alteration involves the replacement of one species with another in the same niche, or the filling of an unoccupied niche, the overall effect on the function of the ecosystem may be neutral or beneficial. Thus, when midwestern prairies were converted to agricultural fields the native prairie chicken was unable to adapt to the altered environment but the introduced ring-necked pheasant, which had become adapted to the agro-ecosystem in Europe (partly, at least, through artificial selection by man), has thrived in the altered landscape. As far as the hunter is concerned the "game bird niche" has been more than adequately filled by the introduction. Too often, however, the introduced species become pests, creating serious environmental problems. Especially grave problems often result with domesticated plants and animals "escape" back to nature and become severe pests because of the absence of both artificial or natural controls. Damage caused by weeds and feral[2] animals to crops, watersheds, forests, and lakes can be extremely costly in terms of diverting energy away from human use. On some of the Hawaiian Islands, feral goats have had a more severe impact on soil, flora, and fauna that has man's plow and bulldozers. Detrimental impact by man on his environment is not confined to industrialized societies nor to the twentieth century. Overgrazing and other types of overexploitation of solar-powered nature have contributed to the downfall of many early civilizations.

Species vary greatly in the rigidity of their niches. Same species may function differently—that is, occupy different niches—in different habitats or geographical regions. The case of the coral, as discussed in the previous section, is probably a good illustration. Man, himself, is another good example. In some regions man's food niche is that of a carnivore (meat eater), while in other regions it is that of a herbivore (plant eater); in mosts cases man is omnivorous (mixed feeder). Man's role in nature, as well as his whole way of life and cultural development can be quite different according to the major energy source on which he depends for food.

Species vary, of course, in the breadth of their niche. Nature has its specialists and its generalists. There are insects, for example, that feed only on one special part of one species of plant, other species of insects may be able to live on dozens of different species of plants. Among the algae there are species that can function either as autotrophs or as heterotrophs; other species are obligate autotrophs only. Although more study is needed, it would seem that the specialists are often more efficient in the use of their resources and, therefore, often become very successful (that is, abundant) when their resources are in ample supply. On the other hand, the specialists are more vulnerable to changes, such as might result from marked environmental or bio-

[2] A feral animal is a domestic animal which becomes readapted to the wild state; its genetic makeup may be different from its original wild ancestor as a result of artificial selection during the domestic period.

logical upheavals or the exhaustion of the resource. Since the niche of nonspecialized species tends to be broader, they may be more adaptable to changes, even though never so locally abundant. Most natural ecosystems seem to have a variety of species, including both specialists and generalists.

DIVERSITY AND STABILITY IN THE ECOSYSTEM

In the preceding section we considered the geographical and qualitative aspects of the distribution of taxa in the ecosystem. Now we will take up the quantitative relations between species and individuals or, more broadly, the relations between the kinds of components and the total number of components. A very characteristic and consistent feature of natural biotic communities is that they contain a *comparatively few species that are common*—that is, represented by large numbers of individuals or a large biomass—and a *comparatively large number of species that are rare at any given locus in time and space.* A tract of hardwood forest, for example, may contain 50 species of trees of which half a dozen or less account for 90 percent of the timber. A tabulation made by an ecology class in its study of a small area of grassland will illustrate the general picture. As shown in Table 2-3, one species comprised 24 percent, 9 species 84 percent, and the remaining 20 species of grasses and herbs only 16 percent of the total stand of vegetation. Each of the latter species accounted for less than 1

Table 2-3. Species Structure of the Vegetation of an Ungrazed Tall-Grass Prairie in Oklahoma

Species	*Percent of stand*[a]
Sorghastrum nutans (Indian grass)	24
Panicum virgatum (Switch grass)	12
Andropogon gerardi (Big bluestem)	9
Silphium laciniatum	9
Desmanthus illinoensis	6
Bouteloua curtipendula (Side-oats grama)	6
Andropogon scoparius (Little bluestem)	6
Helianthus maximiliana (Wild sunflower)	6
Schrankia nuttallii (Sensitive plant)	6
20 additional species (average 0.8 percent each)	16
Total	100

[a] In terms of percent cover of total of 34 percent area coverage of soil surface by the vegetation. Figures are rounded off to nearest whole number. (Data from Rice, *Ecology*, 33:112, 1952, based on 40 one-square-meter quadrat samples taken by an ecology class.)

percent of the community. The few common species in a particular community grouping are often called *dominants,* or *ecological dominants,* if we are thinking of ecological groupings rather than taxonomic ones. Although ecological dominants account for most of the standing crop and community metabolism this does not mean that the rare species are unimportant. In the aggregate they have appreciable impact and determine the amount of diversity in the community as a whole.

Natural communities contain a bewildering number of species, so many in fact that it would be difficult to identify and catalog all the species of plants, animals, and microbes to be found in any large area, as for example, a square mile of forest or a square mile of ocean. Of course, much of the impressive diversity of species observed as we walk around in the environment is due to variations in physical environment that result in mixtures or gradients. However, even if we select small samples from an apparently homogeneous, uniform habitat and restrict ourselves to a limited taxon, the same pattern, as observed in the prairie vegetation, emerges, that is, a few dominant species are associated with many rare ones. A litter of pine needles under a stand composed of one or two species of pine trees is about as uniform a habitat as one can find in nature. If we were to bring in samples of the pine needles and place them on a screen in the top of a funnel under a light bulb, a surprising variety of small animals would crawl out and fall into the bottom of the funnel. Table 2-4 shows what two

Table 2-4. Numbers of Individuals (Adults) of 60 Species of Oribatid Mites Recovered from 215 Samples of Pine Litter[a]

Species	*Number of Specimens*	*Percent of Total*	*Cumulative Percent of Total*
Oppia translamellata	2725	41.2	41.2
Cultroribula juncata	530	8.0	49.2
Tectocepheus velatus	356	5.4	54.6
Galumna sp.	244	3.7	58.3
Scheloribates sp.	208	3.2	61.5
Trhypochthonius americanus	205	3.1	64.6
Peloribates sp.	179	2.7	67.3
Suctobelba palustris	176	2.7	70.0
Zygoribatula sp.	138	2.1	72.1
Remaining 51 species	1828	27.9	100.0
Total	6589	100.0	100.0

[a] Samples were collected from three stands of pine forest (*P. echinata* and *P. Virginiana*), June 29–August 17, east Tennessee (data of Crossley and Bohnsack, *Ecology*, 41:632, 1960).

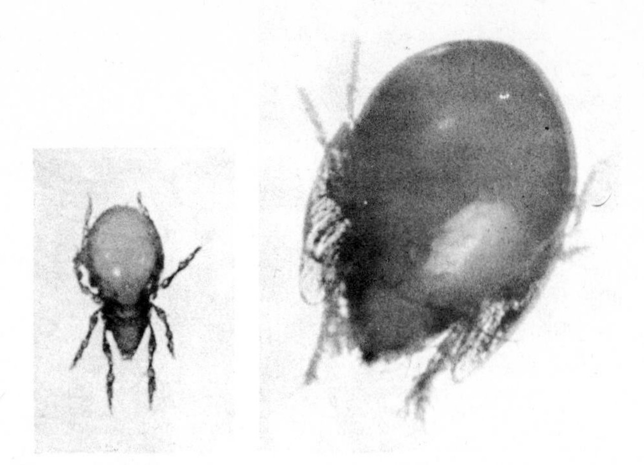

Fig. 2-8 Two species of oribatid mites found in soil and litter, illustratin the diversity of form to be found even within a very limited taxonomic an ecologic compartment. (Photograph by E. F. Menhinick.)

investigators found in the way of oribatid mites in 215 samples of pin litter. Oribatid mites (class Acarina), two of which are shown ir Figure 2-8 to illustrate the variety of form and size, are just one grour of small arthropods (or "microarthropods") that live in forest litte these mites feed on the dead needles, on fungi, or on each other. A shown in Table 2-4, 60 species were identified among 6000 adu individuals recovered from the litter samples. Since the species o immature individuals could not be identified, there may have bee even more species present. Forty-one percent of the adults belonge to only one species and 72 percent of the total to only 9 species, whil 51 species contributed only 28 percent of the individuals. As in th case of the prairie vegetation (Table 2-3), one species was strongl dominant, a small number fairly common, and a large number o species quite rare.

The mite data are treated in another way in Figure 2-9 so as t place the emphasis on the rare end of the species spectrum. The ta black column on the left shows that about half (nearly 30) of th total species were found in less than 10 of the 215 samples; these ar the truly rare species. In contrast, only 5 species occurred in as many a 100 of the samples. When frequency (or number of individuals b classes) is plotted aganist number of species in the manner shown i Figure 2-9, a concave or "hollow" curve is characteristic. The shape o this curve is of great interest to ecologists. An unfavorable limitin factor—for example, a prolonged drought—would tend to make th "hollow" curve more symmetrical (as shown by the dashed lines in Figure 2-9); the number (or percent) of less frequent or rare species would be reduced, and the relative number (or percent) of frequent species would be greater. Analysis of this sort is useful in a practical way when we wish to determine if a man-made limiting factor, such as pollution in a stream or a chronic overdose of insecticide in a forest, is affecting the species structure of the ecosystem. Under very severe stress only a few species survive but their frequency would be high, as in line B in Figure 2-9. In the next section we shall consider more convenient ways of assessing diversity.

The pattern of a few common species associated with many rare species seems to hold regardless of whether we deal with an ecological category, such as "producers" or "herbivores," or with a taxonomic group, for example, Spermatophyta (seed plants) or Acarina. Similar patterns show up when we consider occupations in a city or the kinds of food eaten by a population. The total number of species or whatever unit of structure or function is under consideration is reduced where conditions of existence are severe (as in the arctic) or the geographical isolation is pronounced (as on an island). Size of organism is important; in general, diversity is greater in small organisms than in large. Thus, we would expect to find more kinds of mites than mammals in a

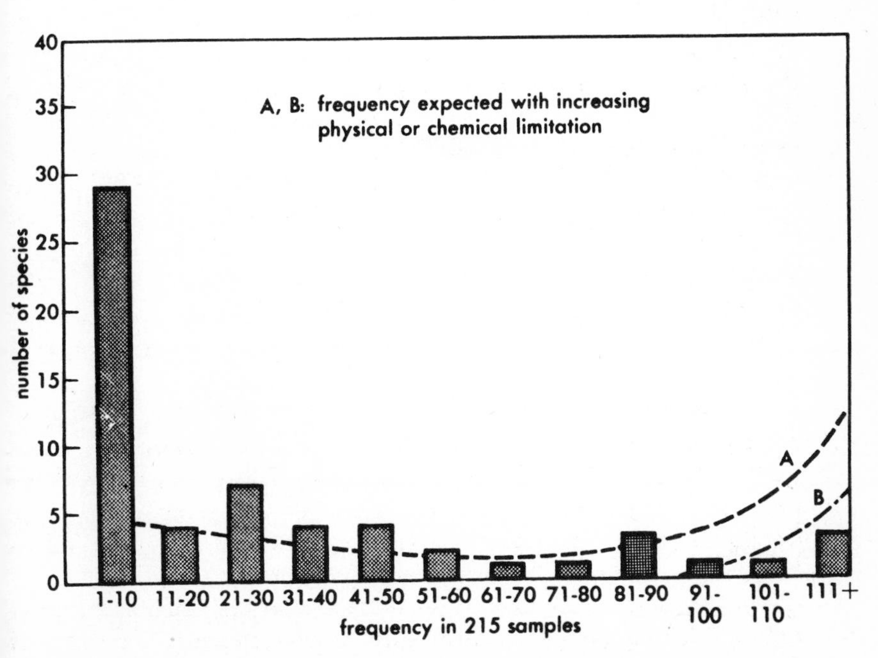

Fig. 2-9 Frequency of occurrence of some 60 species of adult oribatid mites in 215 samples of pine litter in three pine forests in Tennessee.

forest. Man, of course, often exerts strong selective pressure to reduce diversity in order to increase the amount or yield of a desired dominant species, especially in agriculture, forestry, fish and game management, and the like. Even here nature's pattern tends to persist unless the management is very energetic. The species structure of a cultivated grain field in mid-season is displayed in Table 2-5 based on sampling by

Table 2-5. Species Structure of the Vegetation of a Cultivated Millet Field in Georgia

	Percent of Stand[a]
Panicum ramosum (Brown-topped millet)	93
Cyperus sp. (Nut sedge)	5
Amaranthus hybridus (Pigweed)	1
Digitaria sanguinalis (Crabgrass)	0.5
Cassia fasciculata (Partridge pea)	0.2
6 additional species (average 0.05% each)	0.3
	100

[a] In terms of percent dry weight above-ground plants based on 20 quarter-square-meter quadrat samples taken in late July.

another ecology class. Although, as expected, the grain species (millet) constituted a very large portion of the crop, there were 10 other species of herbaceous plants that, in the aggregate, constituted 7 percent of the stand (compare with natural grassland, Table 2-3). In this case no herbicides or other weed control measures were applied. The natural tendency for diversification is such that to maintain a pure culture (monoculture), even in the laboratory, requires a large energy subsidy in the form of mechanical or chemical work. Information obtained from pure cultures may be applied to the better understanding of the nutritional and related niche requirements, but the species must also be studied as it exists in real life, since pure culture conditions never persist in nature. The presence of other species and conditions can completely alter the niche. Thus the pure culture techniques a student learns in elementary bacteriology are inadequate for the study of microbes in soil, water, forest litter, or other natural habitat; we must devise quite different techniques for study "in situ." We have digressed a bit here in order to reemphasize what was brought out in Chapter 1; namely, since both the field and the laboratory approach (or, more broadly the reductionist and holistic approach) have limitations, they must be combined if the complete truth is to be revealed.

DIVERSITY INDICES Although the pattern of many-species-most-of-which-are-rare seems to be almost an ecological "law," the actual number of rare species—and hence the total diversity—is quite variable within and between ecosystems, even though we stick to the same taxonomic or ecological grouping.

A convenient way to express and compare diversity is to calculate diversity indices based on the ratio of parts to the whole, or n_i/N, where n_i is the number or other *importance value* (biomass, productivity, surface coverage, and so on) of each component (species, for example) and N is the total of importance values. The percent of stand or percent of total as shown in Tables 2-3, 2-4, and 2-5 become such ratios when the decimal is shifted two places to the left (24 percent becomes 0.24, for example). Formulas and methods of calculation for two of the most commonly used diversity indices, the Simpson index and the Shannon index, are given in Appendix 2. Ratios for each component are squared and summed to obtain the Simpson index, while each ratio is multiplied by the log of the ratio and the products summed to obtain the Shannon index. Also shown in Appendix 2 is a method of *scaling* the indices so that each has the same numerical range. For the purpose of our discussion indices are scaled 0–1, which means that 0 is the lowest possible diversity (only one kind) and 1, or an approximation of 1, is the maximum diversity for a given number of kinds in that each kind has the same importance value (10 species each with 10 individuals, for example).

The scaled diversity indices for the data in Tables 2-3, 2-4, and 2-5 calculate out as follows (see Appendix 2):

	Simpson Index (0–1)	Shannon Index (scaled, 0–1)
Prairie vegetation	0.8925	0.7861
Mites in pine litter	0.8142	0.6824
Vegetation, millet field	0.1325	0.1356

In a general way we can say that plant diversity in the undisturbed prairie (as based on cover as an importance value) is 89 percent on the Simpson scale and 79 percent on Shannon scale, of the maximum possible diversity in a system of 29 species. The diversity in the mite assemblage is likewise high but, as expected, the diversity of vegetation in the cultivated grain field is very low since there are few kinds and one kind is strongly dominant. The Simpson index is weighted in favor of the common species and the Shannon index in favor of the rare ones (see Appendix 2 for explanation of this statement), with the former scoring higher where there are strong dominants (as in the prairie and mite examples). Because the two indices do convey somewhat different information, together they make a good 2-point profile for diversity assay.

We also see that there are two distinct components that contribute to total diversity. First is the number of kinds, or what we can call the *variety component*; and second is the distribution of relative abundances, or the *evenness component*. The greater the variety (large number of species, for example) and/or the more even are the importance value distribution between the kinds (that is, the lower the dominance), the higher the total or overall diversity. At the research level it is desirable to deal with these components separately in addition to calculating generalized indices.

The three samples (prairie, mites, and grain field) we have discussed fall nicely within the range of diversity that emerges when we compute indices for major functional components of a wide variety of ecosystems ranging from simple to complex and from low-energy to high-energy ones. Ecosystems that are mature in the developmental sense, that is, in steady-state timewise [see Figure 1-2B, and Figure 6-1] and are not subject to severe stress or other disrupting forcing functions generally have a moderately high diversity, in the range of 0.6 to 0.8 of the scale of 0 to 1, but almost never maximum. Ecosystems that are in transient states (Figure 1- 3), or under stress, or tightly managed by man to increase dominance, tend to have low diversities ranging down to 0.0 in one-crop or one-component systems. Diversity indices have proved to be good measures of pollution stress because sewage and industrial wastes almost always reduce the diversity of natural systems into which they are discharged. A measurement of diversity is often better than direct measurement of pollutants, especially where discharge is periodic. For example, waste discharged into a stream in the middle of the night may no longer be detectable by the next day, but the effect on the living community will be evident for some time. For more on diversity and pollution see Patrick (1961), Wilhm and Dorris (1968).

The pattern we seem to see emerging is that the species (or other component) matrix adapts to the strength and variety of energy and material inputs. The strategy of nature is to diversify, but not to the extent of reducing energetic efficiency. Since this is, to some extent at least, contrary to the current strategy of man, we need to inquire into possible reasons for this conflict between man and nature, and to ask to what extent it is necessary or desirable.

Although students of diversity are generally agreed on the proposition that diversity and stability are often directly correlated, it does not necessarily follow that diversity in itself produces stability; it could merely be a result of other stabilizing influences. And we must be careful with the word "stability" since it means different things to different people. The physical scientist generally measures stability in terms of resistance to perturbation; a system is stable if it returns quickly to equilibrium when some outside force knocks it out of equilibrium. The ecologist often thinks of stability in the time-related sense already mentioned; a system is stable if its structure and function remains roughly the same from year to year. The evolutionary biologist thinks in terms of survival; if a population, or system of populations, survives, it is stable no matter how wide the perturbation-related or time-related fluctuations. Despite these differences in viewpoint we can conclude this introductory presentation with several tentative generalizations, and we will cap these off with an example that will interest the citizen of the 1970s, who is deeply concerned with the "energy crisis."

Diversity in systems in general is undeniably a good thing. But as with most "good things" in the real world there can be too much of it as well as too little. We can speculate that the optimum is determined by the energy input into the system (and resource flows coupled with it), since we have documented in this chapter how the kind and level of energy acts as a forcing function to determine the composition as well as the rate of function of ecosystems. When one or a few sources of energy or growth-promoting resources are available in excess to current needs, low diversity has advantages; a concentrated and specialized structure is more efficient in exploiting the bonanza than is a dispersed structure. Putting all the eggs in one or a few baskets, however, makes the system vulnerable when there is a decline or shortage of the major source of the prosperity. Therefore, low diversity, high-energy systems will have a tendency to boom and bust, as we can observe in algal blooms on lakes which receive excess nutrients, or in the rise and fall of yield in one-crop agriculture (see Figure 8-3). Where energy and/or resources are "tight," that is, fully utilized in maintenance, then a higher diversity, as we observe in the prairie grassland, is optimum for the performance of the steady-state system.

In the spirit of model building, as outlined in Chapter 1, the following relationship can be suggested. A diversity on the order of 0.1 (in terms of the 0–1 scale of diversity that we have used for illustration) is characteristic of, and presumably optimum for, the performance of growth systems and perhaps for richly subsidized systems in general, while a diversity in the range of 0.6 to 0.8 is characteristic of steady-state systems and perhaps optimum for unsubsidized solar-powered systems in general. This is a tentative model for testing and not yet proved.

Is this general theory relevant to man's fuel-powered ecosystems? If we consider energy sources as very important "species," then the relative abundances of sources for the United States as of 1970 was something as follows: fossil fuel 95 percent, water power 4 percent, and atomic power 1 percent. A diversity index computed from these ratios would be very low, for example, about 0.08. Mankind has prospered in a material sense and his population has expanded rapidly as a result of his skill in exploiting one major source of energy. Much good for mankind has resulted from this strategy. The problem now is how to avoid the "bust" as this major source declines. Should we now go all out for another dominant source, such as fusion atomic power, to replace fossil fuel in the hopes of continuing the boom, or should we conserve the declining "dominant" and diversify?

It would actually be very difficult for cities and industrialized countries to shift in a short time from one major energy source to another, because the machinery and the whole economic and social structure would have to be redesigned for the new source, and this includes agriculture which is now heavily dependent on oil and gas. Too rapid a shift could result in all kinds of social and economic chaos, disorder, and chain reactions that could degrade the systems beyond our capacity to rebuild. Therefore, the ecologist takes the view that since we are now being forced to power-down and diversify, at least for the next decade or two, then it is prudent to develop a proper strategy for this while we still have enough reserve fuel energy left to make the transition possible. A period of time in a steady-state would provide the incentive to reemphasize human value and quality considerations, and provide sufficient time to determine the best replacement or replacements for fossil fuels. Then mankind would be in a better position to decide whether another boom is possible or desirable. For the immediate future, then, we might envision the development of a distribution pattern of energy sources something as follows: fossil fuel 60 percent, atomic energy 20 percent, solar power, 20 percent, water power 4 percent, and three other sources (geothermal, methane from fermentation, and tidal, for example) 2 percent each. The diversity index for this mix comes to about 0.7, which is within the range that we observe for major components in steady-state natural systems. If we do not at least consider such an option we may soon have no other choice but to suffer the bust, as is "predicted" by the highly publicized study, *Limits of Growth* (Meadows *et al.*, 1972).

SUGGESTED READINGS

References cited

Abrams, Charles. 1965. The use of land in cities. *Sci. Amer.* 213(3): 150–160.

Commoner, Barry. 1971. *The Closing Circle.* New York: Alfred Knopf. (Stresses the need to redesign manufacturing technology to reduce poisonous byproducts and wastes)

MacArthur, Robert. 1972. *Geographical Ecology.* New York: Harper & Row.

Mann, K. H. 1973. Seaweeds: their productivity and strategy for growth. *Science.* 183:975–981.

Meadows, D. H.; D. L. Meadows; J. Randers; and W. W. Behrens. 1972. *The Limits of Growth.* New York: Universe Books.

Odum, H. T. 1971. *Environment, Power and Society.* Chapter 4, pp. 115–134. New York: Wiley-Interscience.

Odum, H. T. and E. P. Odum. 1955. Trophic Structure and productivity of a windward coral reef community on Eniwetok Atoll. *Ecol. Monogr.* 43:331–343.

Patrick, Ruth. 1953. Aquatic organisms as an aid in solving waste disposal problems. *Sewage & Indus. Wastes.* 25:210–214.

Whittaker, R. H. 1969. New concepts of kingdoms of organisms. *Science.* 163:150–160.

Whittaker, R. H.; S. A. Levin; and R. B. Root 1973. Niche, habitat and ecotope. *Amer. Nat.* 107(955):321–338.

Wilhm, J. L. and T. C. Dorris. 1968. Biological parameters for water quality criteria. *Bio-Sci.* 18: 477–481.

The ecosystem as an interdisciplinary concept

Cole, Lamont. 1958. The ecosphere. *Sci. Amer.* 198(4):83–92. (Also included in *Man and the Ecosphere*, a book of Scientific American readings published by W. H. Freeman, San Francisco.)

Duncan, O.D. 1964. Social organization and the ecosystem. In *Handbook of Modern Sociology*, ed. Feris. Chicago: Rand McNally. (A sociologist's view of the ecosystem concept.)

Evans, Francis C. 1956. Ecosystem as the basic unit in ecology. *Science.* 123:1127–1128. (So concludes a biologist!)

Forbes, Stephen A. The lake as a microcosm. A classic essay written in 1887 and reprinted in *Ill. Nat. Hist. Surv. Bull.* 15: 537.

Major, Jack. 1969. Historical development of the ecosystem concept. In *The Ecosystem Concept in Natural Resource Management*, ed. G. M. Van Dyne, pp. 9–22. New York: Academic Press.

Odum, E. P. 1971. *Fundamentals of Ecology*, 3rd ed. Chapter 2, pp. 8–36. Philadelphia: Saunders.

———. 1969. Air-land-water: an ecological whole. *Journ. Soil & Water Conser.* 24:4–7. (Copies available on request to the author.)

———. 1972. Ecosystem theory in relation to man. In *Ecosystem Structure and Function*, ed. J. A. Wiens, pp. 11–24. Oregon State Univ. Press.

Stoddard, D. R. 1965. Geography and the ecological approach. The ecosystem as a geographical principle and method. *Geography*. 50: 242–251. (A geographer's view of the ecosystem concept.)

Vayda, A. P. and Roy Rappoport. 1968. Ecology, cultural and non-cultural. In *Introduction to Cultural Anthropology*, ed. J. Clifton. Boston: Houghton-Mifflin. (Anthropological view of the holism of man and environment.)

The biosphere

Hutchinson, G. E. 1948. On living in the biosphere. *Sci. Monthly*. 67:393–398. Also reprinted in *Readings in Conservation Ecology*, ed., W. W. Cox. Englewood Cliffs, New Jersey: Prentice-Hall. (A classic essay as pertinent today as when it was written 26 years ago.)

———. 1970. The biosphere. Introduction to a special issue of *Sci. Amer.* Vol. 223, No. 3, pp. 44–53. (Also published in book form by W. H. Freeman, San Francisco.)

Cities

Cities. 1965. Special Issue of *Sci. Amer.* Vol. 213, No. 3, September 1965. (Also published in book form, 1973 (with additional articles) by W. H. Freeman, San Francisco.

Detwyler, T. R. and M. G. Marcus. 1972. *Urbanization and Environment; The Physical Geography of the City*. North Scitiuate, Massachusetts: Duxbury Press.

George, Gail and Margaret McKinley. 1974. *Urban Ecology; In Search of an Asphalt Rose*. New York: McGraw-Hill.

Salter, Paul S. 1974. Towards an ecology of the urban environment. In *The Environmental Challenge*, eds. Johnson and Steere, pp. 238–263. New York: Holt, Rinehart and Winston.

Field study of ecosystems

De la Cruz, Armando; E. P. Odum; and June Cooley. 1975. *A Guide to the Study of Ecosystems*. Philadelphia: Saunders (in press).

Odum, E. P. 1957. The ecosystem approach in the teaching of ecology, illustrated with sample class data. *Ecol.* 38:531–535.

Diversity

Odum, E. P. 1971. *Fundamentals of Ecology*, 3rd ed., pp. 143–154. Philadelphia: Saunders.

Whittaker, R. H. 1965. Dominance and diversity in land plant communities. *Science*. 147:250–260.

chapter 3

Energy Flow Within the Ecosystem

In the previous chapter we outlined the gross structure and function of an ecosystem and we classified ecosystems on the basis of type and level of gross energy flow. In terms of the box diagrams of Figure 1-3, we are now ready to consider how the small boxes within the big box (the ecosystem as a whole) are hooked up by flows of energy and exchanges of materials. This will enable us to focus on such interesting and important aspects as the laws of thermodynamics, the concept of gross and net energy, the dynamics of food chains, primary productivity, and food for man.

ENERGY AND MATERIALS Assuming that adapted organisms are present in an area of the biosphere, the number and diversity of organisms and the rate at which they live depends not only on the magnitude of available energy and resources, geographical position, evolutionary history, and other aspects discussed in the previous chapter, but also on the manner in which energy flows through the biological part of the system and on the rate at which materials circulate within the system and/or are exchanged with adjacent systems. It is important to emphasize that nonenergy-yielding materials circulate, but energy does not. Nitrogen, carbon, water, and other materials of which living organisms are composed may circulate many times between living and nonliving entities; that is, any given atom of material may be used over and over again. On the other hand, energy is used once by a given organism or population, is converted into heat; in this degraded form it can no longer power life processes and is soon lost from the ecosystem. The food you ate for breakfast is no longer available to you when it has been respired; you must go to the store and buy more for tomorrow. Likewise, water, paper, and metals in the city can be recycled, but not the energy that powers the city. All living organisms and all machines are alike in that they are kept going by the continuous inflow of energy from the outside.

The one-way flow of energy, as a universal phenomenon, is the result of the operation of the laws of thermodynamics, which are fundamental concepts of physics. The first law states, as you may recall, that energy may be transformed from one type (for example, light) into another (for example, potential energy of food) but is never created or destroyed. The second law of thermodynamics states that no process involving an energy transformation will occur unless there is a degradation of energy from a concentrated form into a dispersed form. Because some energy is always dispersed into unavailable heat energy, no spontaneous transformation (as light to food, for example) can be 100 percent efficient.

The second law of thermodynamics is sometimes known as the *entropy law*; entropy being a measure of disorder in terms of amount of unavailable energy in a closed thermodynamic system. Thus, although energy is neither created nor destroyed, it is degraded when used (transformed) to an unavilable form (dispersed heat). Organisms and ecosystems maintain their highly organized, low-entropy (low-disorder) state by transforming energy from high to low utility states. *We have to be concerned with the quality as well as the quantity of energy,* as alluded to in the previous chapter when we referred to the "dilute" nature of solar energy as compared to [illegible] energy; 1000 cal of sunlight is not the same as 1000

And as already stressed, ecosystems adapt and organize according to both the kind and level of energy. If the quantity or quality of energy flow through a forest or a city is reduced, then the forest or the city literally begins to degrade—or become more disorderly, as it were—unless or until it can reorganize at the lower level.

The interaction of energy and materials in the ecosystem is of primary concern to ecologists. In fact, it may be said that the *one-way flow of energy* and the *circulation of materials* are the two great principles or "laws" of general ecology, since these principles apply equally to all environments and all organisms including man. Furthermore, it is the flow of energy that drives the cycles of materials. To recycle water, nutrients, and so on, requires energy which is not recyclable, a fact not understood by those who think that artificial recycling of man's resources is somehow an instant and free solution to shortages. Like everything else worthwhile in this world, there is an energy cost.

THE SOLAR RADIATION ENVIRONMENT

Organisms at or near the surface of the earth are immersed in a radiation environment consisting of direct downward flowing solar radiation and long-wave heat radiation from nearby surfaces. Both contribute to the climatic regime that determines "conditions of existence," as noted in the previous chapter, but only a small fraction of the direct solar component can be converted by photosynthesis to provide food energy for the biotic components of the ecosystem. Extraterrestrial sunlight reaches the biosphere at a rate of 2 g-cal/cm^2/min. This quantity is known as the *solar constant*. Since the sun shines only for part of the day at any location, the amount coming in on a day or year basis is about half, more or less. On a square-meter basis this comes to about 14,400 kcal/day or 5.25 million kcal/year. This large flow is reduced exponentially as it passes through clouds, water vapor, and other gases of the atmosphere, so the amount actually reaching the autotrophic layer of ecosystems is on the order of 1.0 to 2.0 million kcal/m^{-2} year^{-1}, depending on latitude, cloud cover, and so on. Of this about half is absorbed by a well-stocked green layer and 1–5 percent of this converted to organic matter that structures and operates the solar-powered ecosystem.

The sequence of energy flow that we have just described, including further transfers to animals and man, is shown in the diagrammatic model of Figure 3-1. Quantities shown are much rounded-off averages that are appropriate for a north temperate latitude such as mid-continent North America. On an annual basis we see how rapidly solar energy is lost into the heat sink (i, ii, and so on in Figure 3-1B) as it passes through the atmosphere and the green belt. The organic food that plants are able to produce from sunlight is partly used by the plants themselves for their own maintenance and growth (with appropriate heat loss) and is partly passed on to the heterotrophs. In the diagram C_1 represents the primary consumer or herbivore level and C_2 the secondary consumer or carnivore level. In the plant-animal portion of the energy flow chain about 80 to 90 percent of the energy is lost with each step, or to put it another way, only 10–20 percent can be transferred to the next level. Thus, out of the millions of calories of solar energy coming into a column with a square-meter base, only a few hundreds are left to nourish a meat-eating animal, or man. Two sets of figures are shown in the right-hand portion of Figure 3-1: (1) along the top of the line averages for biosphere as a whole, and (2) in

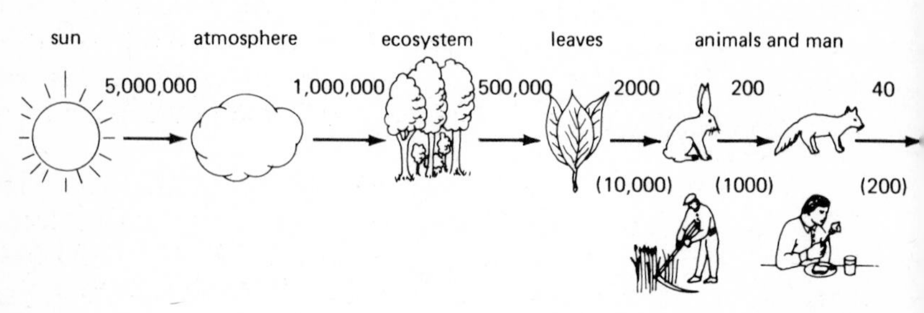

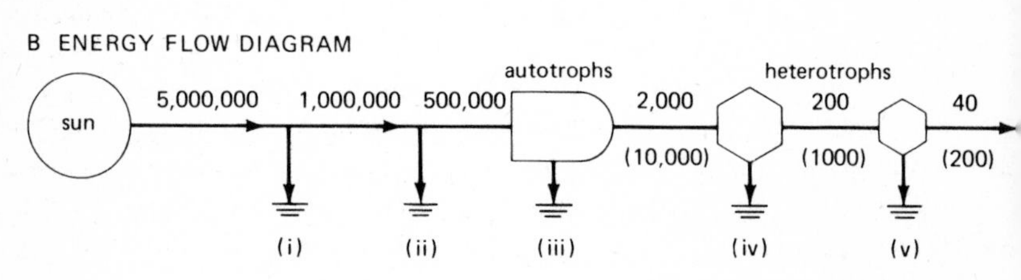

Fig. 3-1 Solar-energy flow in kcal/m² per year, as shown in a pictorial diagram (A) and in a more formalized flow diagram (B). The heat sink symbol (⏚) in (B) shows where energy is lost in transformation. Five losses where useful work is done are as follows: (i) Attenuation of extraterrestrial sun energy in heating the atmosphere and driving hydrological cycles and weather systems. (ii) Attenuation of sun energy to warm the ecosystem and drive its internal water and mineral cycles. (iii) Energy loss in conversion of sun energy to plant matter. (iv) Energy loss in conversion of plants to herbivores (primary consumers (C_1). (v) Energy loss in transfer from prmiary to secondary consumers (C_2). The figures in parentheses in the biological part of the energy chain represent levels for subsidized ecosystems (see Table 3-1).

parentheses below the line 10 times these figures for the favored solar-powered ecosystem that receive supplemental energy (compare Table 2-1).

It is important to note that useful work is accomplished at each transfer, not just in the biological part but all along the chain. Thus, although we cannot eat much of it, or use it directly to run our machines, all of the incoming solar radiation is vital to the operation of the biosphere. For example, the dissipation of solar radiation (A and B, Figure 3-1) as it passes into the atmosphere, the seas, and the green belts warms the biosphere of life-tolerable levels, drives the hydrological cycle (see Figure 4-6), and powers weather systems. So delicate are the heat and other energy balances of the earth that meteorological models now show that only very small changes in the solar constant, or in the turbidity of the atmosphere (which would let more or less energy reach the surface of the earth) are needed to change the world's climates. Just a little bit of decrease in heat brings on an ice age, while a small increase brings on a tropical era, with a melting of all the polar ice raising the sea level to flood large areas of present continents. (Good-by New York and most of the world's large coastal cities!)

In Figure 3-1 we introduce a symbolic "energy language" which has been developed by Howard T. Odum to facilitate communication between physical scientists and engineers on the one hand, and biologists and social scientists on the other (see H. T. Odum, 1971, page 38).[1] In this and subsequent diagrams of this type circles signify energy sources, the sun in this instance. The heat sink symbol (i through v in Figure 3-1B), shows where energy is lost in transformation from one form to another as required by the second law of thermodynamics. The heat sink symbol resembles an electrical ground symbol, but is one-way (as indicated by downward directed arrow). The bullet-shaped symbol represents an autotrophic system (or more broadly a unit capable of receiving pure wave energy, such as light, and producing an energy-activated state, such as food, which can be deactivated to pass energy on to another step in a chain of energy flow). The hexagonal symbol represents a heterotrophic unit, or more broadly a self-maintaining component that is capable of receiving, storing, and feeding back energy received from an autotrophic, another heterotrophic, or another concentrated potential energy source. Additional symbols will be introduced in subsequent diagrams.

The spectral, that is, the wavelength, distribution of sunlight is also altered as it passes through atmosphere, clouds, water, and vegetation. The ozone belt of the upper atmosphere selectively absorbs the lethal short-wave ultraviolet radiation so that only about 10 percent

[1] A similar sign language is used by Jay Forrester in his models of "world dynamics" (1971) that were the basis for the Meadow's *Limits of Growth* study.

reaches the earth's surface on a clear day. Visible radiation (medium wavelength) on which photosynthesis depends is least attenuated as it passes through clouds and water, which means that photosynthesis can continue on cloudy days, and at some depth in lakes and the sea (if they are not too turbid). Green plants efficiently absorb the blue and visible red light that is most useful in photosynthesis and reject, as it were, the near infrared heat waves and thus avoid overheating. The long-wave infrared radiation, in general, which makes up the bulk of solar energy, is absorbed and reradiated as heat in a complex manner by atmosphere, clouds, and various natural and man-made objects and surfaces. For more on these aspects, see Gates (1963). Just because the world's green belts convert only a small percentage of incoming solar energy to food energy, does not mean that they are inefficient; actually, photosynthesis is a very efficient process for tapping that small portion of sunlight that can readily be converted to high utility potential energy of organic matter.

FOOD CHAINS AND TROPHIC LEVELS

The transfer of food energy from the source in plants through a series of organisms with repeated stages of eating and being eaten is known as the *food chain*. In complex natural communities, organisms whose food is obtained from plants by the same number of steps are said to belong to the same *trophic level* (troph = nourishment, the same root as in autotrophic and heterotrophic). Thus, green plants occupy the first trophic level (the producer level); plant eaters (herbivores, and so on) the second level (the primary consumer level); carnivores that eat the herbivores the third level (secondary consumers), and perhaps even a fourth level (tertiary consumers). It should be emphasized that this trophic classification is one of function, and not of species as such; a given species population may occupy one, or more than one, trophic level according to the source of energy actually assimilated. We have already called attention to certain algae that may depend in part on their own food and in part on food made by other algae; or the populations of men who utilize food from both plant and animal sources.

Figure 3-2 is an energy flow diagram of a food chain. This is a more detailed rendition of the last three modules in Figure 3-1. In this model we introduce the biological terminology used to describe community metabolism—and the quantities shown are on a daily, rather than annual, basis. The boxes represent the population mass or biomass, and the pipes depict the flow of energy between the living units. As already indicated, about half of the average sunlight impinging upon

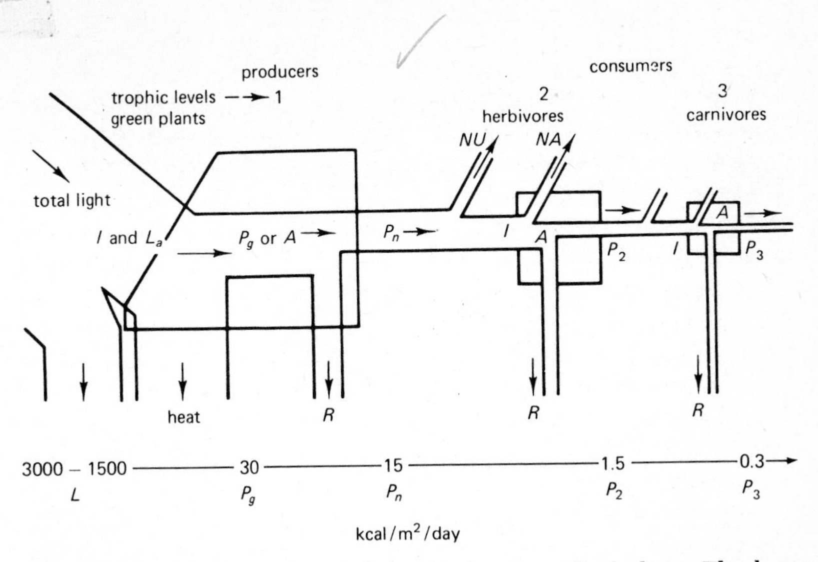

Fig. 3-2 A simplified diagram of energy flow in a food chain. The boxes represent the standing crop of organisms (1: producers or autotrophs; 2: primary consumers or herbivores; 3: secondary consumers or carnivores) and the pipes represent the flow of energy through the biotic community. L = total light; L_a = absorbed light; P_g = gross primary production; P_n = net primary production; P = secondary production at second (P_2) and third (P_3) trophic levels; I = energy intake; A = assimilated energy; NA = nonassimilated energy; NU = unused energy (stored or exported); R = respiratory energy loss. The chain of figures along the lower margin of the diagram indicates the order of magnitude expected at each successive transfer starting with 3000 kcal of incident light per m² per day.

green plants (that is, producers) is absorbed by leaves and about 1 to 5 percent is converted into food energy by productive vegetation. The total assimilation rate of producers in an ecosystem is designated as *primary production or primary productivity* (P_g or A in Figure 3-2). It is the total amount of organic matter fixed, including that used up by plant respiration during the measurement period. *Net primary productivity* (P_n) is the organic matter stored in plant tissues in excess of respiration during the period of measurement. Net production represents food potentially available to heterotrophs. When plants are growing rapidly under favorable light and temperature conditions, plant respiration may require as little as 10 percent of gross production so that net production may be 90 percent of gross. However, under most conditions in nature net production is a smaller percentage of gross, usually about 50 percent, as shown in the lower line of Figure 3-2.

As shown in Figure 3-2 part of net primary production may be stored or exported, and part may become an energy source for heterotrophs. Later in the chapter we will consider how the two alternative flows, *NU* and *I*, are apportioned within different kinds of ecosystems. Some portion of food ingested by consumers is usually not digestible or assimilable, so some of the energy is likely to be egested unused (*NA* in the diagram); this component may be stored, exported, or consumed by microorganisms or other heterotrophs. Consumers as well as producers must respire (*R*) a large part of the energy assimilated (*A*) so as to maintain structure and function (represented by the boxes in the diagram). Respiratory degradation of energy "pumps out" disorder, that is, reduces entropy as is required to maintain a high level of organization. The *R* flows in Figure 3-2 represent heat losses from biological components that were shown by the more generalized heat sink symbol in Figure 3-1. Assimilated energy not respired is available for production (P_n, P_2, P_3, . . .) which can take the form of growth of new tissue and reproduction and growth of individuals (population growth). Production at heterotrophic levels is often known as *secondary production* to distinguish it from the primary production of plants. As pointed out in Chapter 2, any auxiliary energy or nutrients that reduce the cost of maintenance (respiration) enhances the rate of production. If the autotrophic level is subsidized, the increased production may be passed along the chain; if a consumer level is augmented, the "downstream" links will be mostly affected, although secondary production can "loop back" or be "fed back" upstream so that even plants might benefit from organic matter not used by animals.

Roughly speaking, the reduction of available energy with each link in the food chain (as required by the second law of thermodynamics) is about two orders of magnitude at the first or primary trophic level, and about one order of magnitude thereafter. By order of magnitude we mean by a factor of 10. If an average of 1500 kcal of light energy were absorbed by green plants per square meter per day, we might expect 15 to end up at net plant production, 1.5 to be reconstituted as primary consumers (herbivores), and 0.3 as secondary consumers (carnivores)—provided, of course, that there are adapted organisms present that can fully utilize these resources. Efficiencies in terms of percent energy transfer are thus on the order of 1 percent at the first level and 10–20 percent at the heterotrophic levels. Since meat is generally of higher nutritional quality, percent transfer tends to be higher at meat-eating levels. Since so much energy is lost at each transfer the amount of food remaining after two or three successive transfers is so small that few organisms could be supported if they had to depend entirely on food available at the end of a long food chain. For all practical purposes, then, the food chain is limited to three or four "links." The shorter the food chain, or the nearer the organism to the beginning of the food chain, the greater the available food energy.

About ten times more people can be supported by 100 acres of corn if they function as primary rather than secondary consumers, that is, if they eat the corn directly rather than feed it to animals and then eat meat. If you prefer to eat meat then you must plan not to let the population become so dense as to preclude the meat-eating option! In debating man's "world food problem" we must consider the quality as well as the amount that might be obtained from a given area of land or water.

It should be emphasized that the scheme in Figure 3-2 is a model useful for making comparisons with the real thing. Somewhat larger, and frequently quite a bit smaller, percentages may actually be involved under different conditions. Much remains to be learned, not only about the orders of magnitude in different ecosystems but also about the upper limits. Since the efficiencies of transfer seem low in terms of man-made machines, man has often thought that he could improve on nature by increasing the percent of transfer of light to food, and food to consumer. However, when we consider the low quality of available energy, the fact that organisms unlike machines are self-maintaining, and the need for storage and diversity for future survival (as discussed in Chapter 2), then it turns out that nature's efficiencies are just about optimum. (For further discussion of this important point, see Odum and Pinkerton, 1955; H. T. Odum, 1971, Chapter 3.)

At this point it would be well to make clear a few points about units, ratios, and efficiencies. Too often our thinking is clouded by people who, unintentionally or intentionally, use units that are not comparable in forming ratios and percentages. Biomass varies in energy value, both quantitatively and qualitatively. Plant biomass runs about 4.5 kcal per ash-free dry gram, and animal close to 5.5. Where energy is being stored, as in seeds or in bodies of migrating or hibernating animals; the values approach 7 or 8 kcal. If plant matter is mostly cellulose and lignin (wood), its energy (4 kcal/g) is unavailable to most animals. The main point to remember about ratios and percentages when used as estimates of efficiency is that they should be *dimensionless;* that is, the same unit should be used for both the denominator and numerator of the ratio. Thus, it should be calories/calories not calories/grams. Otherwise comparisons may not be valid. For example, poultry or catfish farmers may tell you they are getting one pound of meat for every two pounds of feed, implying a 50 percent efficiency of transfer. However, since the food is highly concentrated, dry material worth 5 or more kcal/g, and the meat is "wet weight," worth only about 2 kcal/g, then the real energy out/energy in efficiency is more on the order of 20 percent.

THE RELATIONSHIP BETWEEN ENERGY FLOW AND THE STANDING CROP

As already indicated, the boxes in Figure 3-2 represent the biomass of the standing crop of organisms functioning at the trophic level indicated. The relationship between the "boxes" and the "pipes"—that is, between standing crops and the energy flows *P*, *A*, or *I*—is of great interest and importance. As we have seen, the energy flow must always decrease with each successive trophic level. Likewise, in many situations, the standing crop also decreases (as shown in Figure 3-2). However, standing crop biomass is much influenced by the size of the individual organisms making up the trophic group in question. In general, the smaller the organism the greater the rate of metabolism per gram of weight. This trend is often known as the *inverse size–metabolic rate* "law," and has already been noted in the previous chapter. Consequently, if the producers of an ecosystem are composed largely of very small organisms, and the consumers are large, the standing crop biomass of consumers may be greater than that of the producers even though, of course, the energy flow of the latter must average greater (assuming that food used by consumers is not being "imported" from another ecosystem). Such a situation often exists in marine environments where the water is moderately deep: bottom-dwelling invertebrate consumers (clams, crustaceans, echinoderms, and so on) and fish often outweigh the microscopic phytoplankton on which they depend. By harvesting at frequent intervals, man (as well as the clam) may obtain as much food (net production) from mass cultures of small algae as he obtains from a grain crop harvested after a long interval of time. However, the standing crop of algae at any one time would be much less than that of a mature grain crop.

To reiterate, standing crop is a measure of the amount of living material present at a particular time. Productivity is a *rate* to be expressed as energy flow per unit area per unit time. As indicated by the examples, these two quantities should not be confused; the relationship between the two depends on the kind of organisms involved. Standing crop can be used as an index of productivity only if production accumulates unused, as in a crop where harvest is deferred until the end of the season. If growth is used as fast as it is produced (as in a grazed pasture), then standing crop cannot be used to estimate productivity.

PRODUCTION AND UTILIZATION RATES

The relationship between gross production (P_g) and total community respiration (the sum of all *R*'s in Figure 3-2) is important in the understanding of the total function of the ecosystem and in predicting future events. One kind of ecological "steady-state" exists if the annual production of organic matters equals total consumption ($P/R=1$) and if exports and imports of organic matter are either nonexistent or equal. In a mature tropical rain forest the balance may be almost a day-by-day affair, whereas in mature temperature forests an autotrophic regime in summer is balanced by a heterotrophic regime in winter. Another type of steady-state exists if gross production plus imports equal total respiration, as in some types of stream ecosystems, or if gross production equals respiration plus exports, as in stable agriculture.

Seasonal fluctuations and annual shifts related to short-term meteorological or other cycles in the physical environment occur in almost all ecosystems, but the overall structure and species composition of steady-state communities tend to remain the same, although it is not yet certain that this is always true. If primary production and

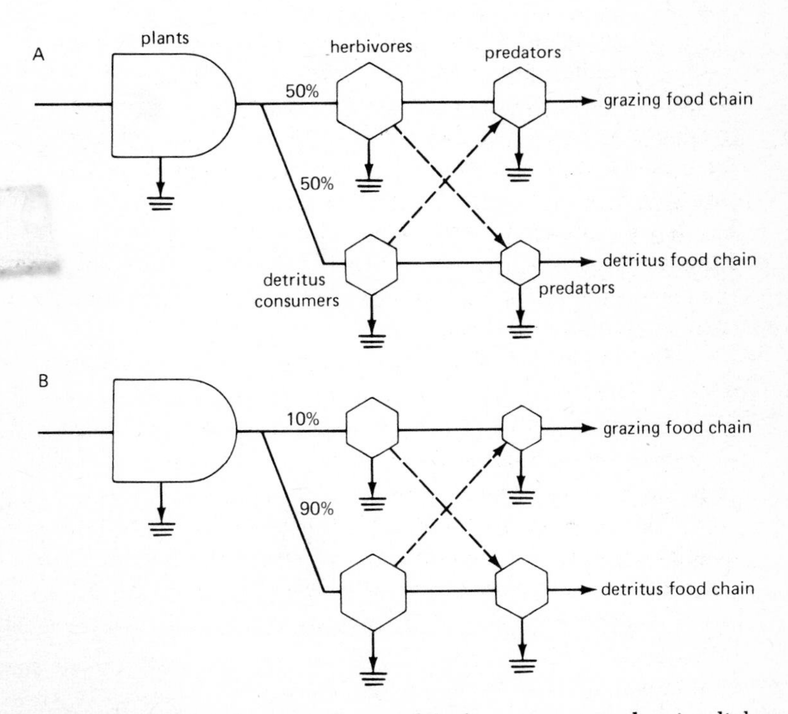

Fig. 3-3 The Y-shaped energy flow model of an ecosystem showing linkage of two major food chains; the grazing food chain and the detritus food chain. Diagram (A) represents an ecosystem, such as a grazed pasture, with a large proportion of energy (50%) flowing through the grazing pathway. Diagram (B) represents an ecosystem, such as a mature forest, with most energy flowing along the detritus pathway.

heterotrophic utilization are not equal (P/R greater or less than 1), with the result that organic matter either accumulates or is depleted, we may expect the community to change by the process of *ecological succession.* Succession may proceed either from an extremely autotrophic condition ($P>R$) or from the extremely heterotrophic condition ($P<R$) toward a steady-state condition in which P equals R. Organic development in a new pond, or the development of a forest on a fallow field are examples of the first kind of succession. In these situations the kinds of organisms change rapidly from year to year and organic matter accumulates. Changes in a stream polluted with a large amount of organic sewage is an example of the other type of succession, in which organic matter is used up faster than it is produced. Ecological succession will be discussed in greater detail in Chapter 6.

The ratio of biomass energy to rate of energy flow is an important property of ecosystems as is the P/R ratio. In the ecosystem the ratio of total community respiration to total community biomass (R/B) can be considered to be a thermodynamic order function, for reasons already made clear. The larger the biomass the larger the respiration, of course, but if the size of the biomass units is large and the structure diverse and well ordered, the respiratory maintenance cost *per unit of biomass* can be decreased. Nature's strategy seems to be to reduce the R/B ratio (or increase the B/R efficiency if you prefer) while man's strategy has tended to be the opposite, since he has been preoccupied with harvesting as much as possible and leaving as little structure and diversity on the landscape as possible. As discussed in Chapter 2, this latter strategy can lead to boom and bust oscillations.

GRAZING AND DETRITUS FOOD CHAINS: THE Y-SHAPED ENERGY-FLOW MODEL

As was discussed in the chapter on ecosystems, dividing the heterotrophs into large and small categories—that is, macroconsumers and decomposers—is arbitrary in terms of function but convenient in terms of analysis and study. In the simplified diagram of Figure 3-2, the bacteria and fungi that decompose plant tissues and stored plant food would be placed in the primary consumer box along with the herbivorous animals; likewise, microorganisms decomposing animal remains would go along with the secondary consumers or carnivores. However, since there is usually a considerable time lag between direct consumption of living plants and animals, and the ultimate utilization of dead organic matter, not to mention the metabolic differences between animals and microorganisms (as emphasized in Chapter 2), a much more realistic energy-flow model is obtained if the decomposers are placed in a separate box, and the flow of net production energy is divided into two chains, as shown in Figure 3-3.

In nearly all ecosystems some of the net production is consumed as living plant material and some is consumed later as dead plant material. We can conveniently designate these primary consumers that eat living plants as *grazing herbivores,* whether they be large animals such as cattle or deer, or small animals such as zooplankton. The flow through grazers can be designated as the *grazing food chain* (Figure 3-3). Likewise, consumers of dead matter can be conveniently designated as *detritus consumers* and the flow along this route as the *detritus food chain,* as shown by the lower pathway in the Y-shaped diagrams of Figure 3-3. Two types of organisms consume detritus: (1) small detritus-feeding animals, such as the soil mites or millipedes on land and various worms and mollusks in water; and (2) the bacteria and fungi of decay. These two groups are so intimately associated that it is often difficult to determine their relative effect on the breakdown of the original primary production. In many cases the two seem to be in partnership, since the reduction of large pieces to small pieces by animals makes the material more available to microorganisms, which in turn provide food for the small animals.

In Figure 3-3 a grazed pasture ecosystem and a temperate, deciduous forest ecosystem, in which the detritus food chain predominates, are compared. In a heavily grazed pasture or grassland, 50 percent or more of the annual net production may pass down the grazing herbivore energy-flow path. Since not all of the food eaten by grazers is actually assimilated, some (in feces, for example) is diverted to the decomposer route; thus, the impact of the grazer on the community depends on the amount of plant material removed from the standing crop as well as on the amount of energy in the food that is utilized. Obviously, there must be a limit to direct grazing, since too rapid a removal will kill the producers or greatly reduce their future productive capacity. Range managers (applied ecologists concerned with the wise use of grasslands by man) generally work on the basis that not more than 50 percent of the forage production should be removed by cattle in a season, which means that somewhat less can be assimilated by the cows. In natural communities there seem to be a number of feedback mechanisms that keep grazing herbivores under control. These mechanisms will be described in a later chapter. When man takes over the control of natural grassland communities, he too often fails to regulate his cattle, sheep, and goats until they have gone too far.

In contrast to the grazed pasture type of ecosystem, less than 10 percent of the net production of a temperate forest is consumed by grazing herbivores, mostly insects in this case; at least 90 percent follows the detritus path of energy flow. One ingenious investigator (Bray, 1961) collected autumn leaves as they fell in a deciduous forest and carefully measured "bites" taken out by the grazing insects; he came up with an estimate of 7 percent of the annual crop of leaves consumed by grazers in a season. The estimate, of course, did not include potential energy removed by sucking insects that feed on juices of the plants. Nevertheless, much of the net production in a forest clearly goes into the detritus box (dead leaves, twigs, and so on), resulting in an abundance of consumers associated with the litter and soil. Occasionally, of course, something goes wrong and insects strip all of the leaves from trees. When this happens ecologists are often unable to determine the cause, since so little is known about the normal complex regulatory mechanisms that prevent such a "grazing cancer" in 99 out of 100 cases. In some cases it is clear that the breakdown is the result of pests or other stress by man.

When we examine aquatic communities we find the same contrasts as in terrestrial communities; in some the grazing pathway predominates and in others most energy flows along the detritus pathway. In Long Island Sound, which has been studied in detail by Gordon Riley and colleagues (see Riley, 1956), from one half to three fourths of the phytoplankton production is grazed by zooplankton. In contrast, less than 10 percent of the marsh grass in Georgia estuaries is grazed, mostly by small insects, in this case (see Teal, 1962; E. P. Odum and de la Cruz, 1967); the bulk of the grass falls into the water, is quickly broken up into detritus particles, which are then enriched (that is, protein content increased) by microorganisms and consumed by small animals which in turn support a two-link predator chain consisting of small fish that are fed upon by large fish or other predators.

Very frequently a detritus food chain links two quite different ecosystems, with primary production occurring in one of the linked systems and secondary production (heterotrophic utilization) in the other. For example, practically all of the aquatic organisms living in a woodland stream are supported on a food chain that begins with dead leaves and other detritus that falls in, is blown in, or washes in from the forested watershed. Had it ever occurred to you that the beautiful trout you caught in the mountain stream is from a population supported by dead leaves? Figure 3-4 shows a pictorial model of a detritus food chain based on mangrove leaves as worked out in considerable detail by W. E. Odum and Heald (1972). Red mangroves form forests in shallow water around the southern end of Florida, and south into Central and South America. Very little of the leaves are directly grazed; most fall into the warm water and are transported by tides and currents over large areas of coastal waters. Leaf fragments are acted upon by saprotrophs (bacteria and fungi) and microscopic algae add enrichment to the detritus mix which is eaten and re-eaten (coprophagey) by a key group of small detritus consumers (detritivores), as shown in Figure 3-4. Note that the eating and re-eating of particles and feces is *not* a recycling of energy (which cannot occur, as previously emphasized); it is just that the substrate is resistant and may be passed through animal guts several times before the original detrital

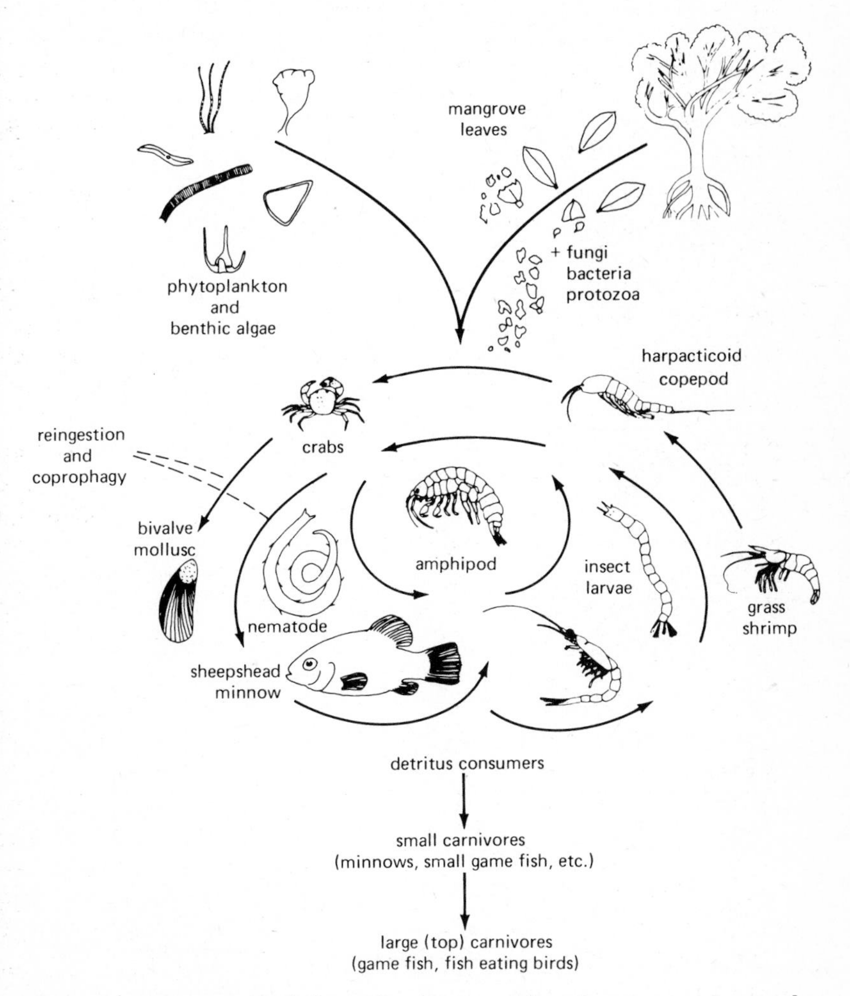

Fig. 3-4 A detritus food chain that begins with red mangrove leaves that fall into shallow subtropical estuarine waters. The decaying detritus particles are enriched by microorganisms and provide food for a key group of detritus consumers which in turn are food for fish. (After William E. Odum, reproduced from E. P. Odum, *Fundamentals of Ecology*, 3rd ed. W. B. Saunders, 1971).

material is all used up. As the material passes through the animal the microorganisms are digested off and the substrate egested to be recolonized again by another crop of microorganisms. In other words, the leaf substrate is the energy source for the decomposer microorganisms, which in turn are the principal energy source (food) for the detrital animals, which, in turn, support the fish. This study demonstrated that red mangroves, which were previously considered to have no economic value, make an important contribution to coastal fisheries. Finally, the model of Figure 3-4 could serve equally well for the salt marsh estuary and the woodland stream detrital systems; only the names of organisms need be changed

To summarize, it is convenient, even if a bit arbitrary, to divide primary consumption into two broad energy flows, and therefore to think in terms of two rather different food chains. The vegetation-rabbit-fox, or phytoplankton-zooplankton-whale, or grass-cow-man sequences are the direct, relatively simple, food chains of classical ecology. The detrital food chain is more complex, less understood (especially the microbial part), and in many ecosystems the more important. Both energy-flow routes exist in nearly all ecosystems but in widely different proportions. It is obvious that any ecosystem that accumulates and stores energy, as in a large forest, must function as a detrital, or delayed consumption system, since if plant production is grazed as fast as it is produced, there can be no accumulation of biomass. As already noted, such a storage system (or linkage of storage systems) may be more stable in terms of resistance to weather and other perturbations. As with any situation, extremes clearly become detrimental, but in the case of ecosystems we are not always aware of what is extreme. For example, overgrazing by definition is detrimental, but what constitutes overgrazing in different ecosystems needs to be clearly defined in terms of energy flow. Likewise, "undergrazing" also can be detrimental. In the complete absence of direct consumption of living plants, detritus may pile up and the release and recycling of minerals may be delayed. Microorganisms, alone, may not be able to break down the bodies of plants and release minerals fast enough if the material becomes dry or anaerobic. In terrestrial nature, fire often acts as a consumer where there is a pileup of undecayed plant material. As will be noted in Chapter 4, there are types of ecosystems that *require* fire for their preservation. In some types of ecosystems man has learned to use fire in a controlled manner to improve the production of timber. Again, the whole ecosystem must be considered; depending on the type of ecosystem and climate, fire and grazers in moderation may be good things; in excess they are bad.

THE AREA-BASED CHLOROPHYLL MODEL In outlining the pond study in Chapter 2, it was suggested that the amount of chlorophyll per unit of area might be interesting as an index for the entire ecosystem. Let us now consider the distribution of this magic green pigment which keys the enzyme system that enables plants to convert light into food. Figure 3-5 shows the amount of chlorophyll to be expected per square meter in four types of ecosystems. That all of nature might be included (from the viewpoint of a simplified model) in four basic types of light-adapted communities has

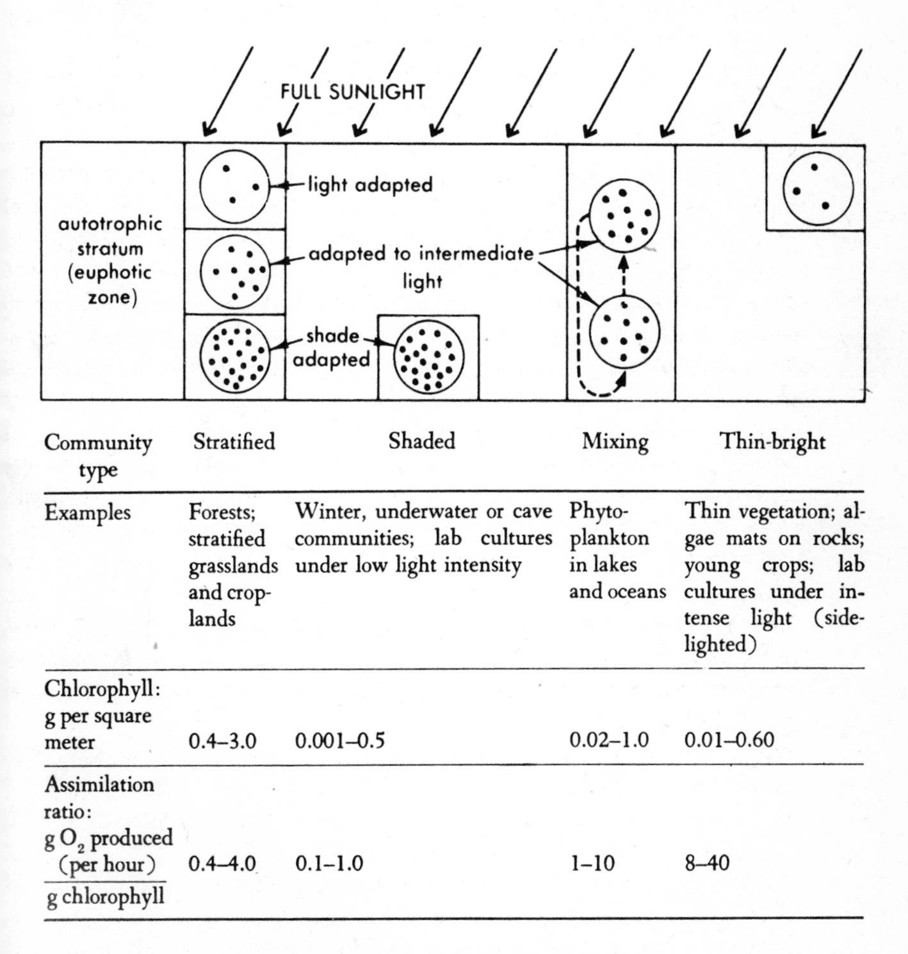

Community type	Stratified	Shaded	Mixing	Thin-bright
Examples	Forests; stratified grasslands and croplands	Winter, underwater or cave communities; lab cultures under low light intensity	Phytoplankton in lakes and oceans	Thin vegetation; algae mats on rocks; young crops; lab cultures under intense light (side-lighted)
Chlorophyll: g per square meter	0.4–3.0	0.001–0.5	0.02–1.0	0.01–0.60
Assimilation ratio: $\frac{\text{g } O_2 \text{ produced (per hour)}}{\text{g chlorophyll}}$	0.4–4.0	0.1–1.0	1–10	8–40

Fig. 3-5 The amounts of chlorophyll to be expected in a square meter of four types of communities. The relation of area-based chlorophyll and photosynthetic rate is also indicated by the ratio between chlorophyll and oxygen production. (After H. T. Odum.)

been proposed by H. T. Odum, W. McConnell, and W. Abbott (1958). The dots in the diagrams indicate relative concentration of chlorophyll per cell (or per biomass). The relation of total chlorophyll and the photosynthetic rate is also indicated by the grams of organic matter produced per hour by a gram of chlorophyll under the light to which the system is adapted, as shown in the bottom row of numbers. This ratio is often called the *assimilation ratio*. As may be seen in the figure, shade-adapted plants or plant parts tend to have a higher concentration of chlorophyll than light-adapted plants or plant parts, thus enabling them to trap and convert as much scarce light as possible. Consequently, efficiency of light utilization is high in shaded systems, but the photosynthetic yield and the assimilation ratio is low. Algae cultures grown in weak light in the laboratory often become shade adapted. The high efficiency of such shaded systems has been sometimes mistakenly projected to full sunlight condition by those who are enthusiastic about the possibilities of feeding mankind from mass cultures of algae; when light intensity is increased in order to obtain a good yield the efficiency goes down, as with other kinds of plants.

Total chlorophyll is highest in stratified communities such as forests and is generally higher on land than in water. For a given light-adapted system the chlorophyll in the photosynthetic zone self-adjusts to nutrients and other limiting factors. Consequently, if the assimilation ratio and the available light are known, total photosynthesis can be estimated by the relatively simple procedure of measuring area-based chlorophyll. Such estimations have been especially useful at sea where direct measurement of rates of production are difficult.

Two Kinds of Photosynthesis

As you may recall, the basic photosynthetic process is chemically an oxidation-reduction-reaction which can be written as follows:

$$\underset{\text{carbon dioxide}}{CO_2} + \underset{\text{water}}{2H_2O} \xrightarrow{\text{light}} \underset{\text{carbohydrate}}{(CH_2O)} + \underset{\text{water}}{H_2O} + \underset{\text{gaseous oxygen}}{O_2}$$

the oxidation being

$$2H_2O \longrightarrow 4H + O_2$$

and the reduction being

$$4H + CO_2 \longrightarrow (CH_2O) + H_2O$$

A recent dsicovery that plants differ in biochemical pathways for the carbon dioxide reduction has important ecological implications. In most plants carbon dioxide fixation follows a C_3 *pentose phosphate cycle* or *Calvin cycle* which for many years has been "the" accepted scheme for photosynthesis. Then, in the late 1960s several plant physiologists, notably Hatch and Slack of Australia, contributed to the discovery that certain plants reduce carbon dioxide in a different manner according to a C_4 *dicarboxylic acid cycle.* It was also found that the latter plants have a different arrangement of chloroplasts within the leaf and, more important, they respond differently to light, temperature, and water. For the purposes of our discussion of the ecological implication of these discoveries we will designate the two types of plants as *C_3 plants* and *C_4 plants.*

Figure 3-6 contrasts the response to light and temperature of C_3 and C_4 plants. C_3 plants tend to peak in photosynthetic rate (per unit of leaf surface) at moderate light intensities and temperatures, and to be inhibited by high temperatures and the intensity of full sunlight. In

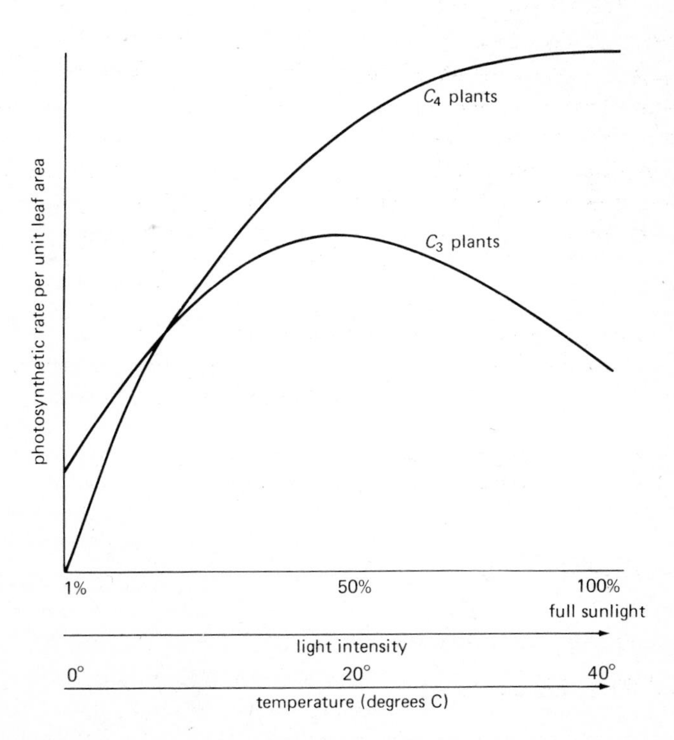

Fig. 3-6 Comparative photosynthetic response of C_3 and C_4 plants to increasing light intensity and temperature. See text for explanation.

contrast, C_4 plants are adapted to high light and high temperature. They are also more efficient in water use, requiring 250–350 g of water to produce 1 g of dry matter, as compared to the 400–1000 requirement for C_3 plants. These are also not inhibited by high oxygen concentrations as are C_3 plants. One reason C_4 plants are more efficient at the high end of the light-temperature scale is that there is little or no "photorespiration," which is to say that the plant's net production is not burned up as light intensity increases. Although it has not yet been checked out thoroughly, C_4 plants seem to be more resistant grazing by insects. Species with C_4 type photosynthesis are especially numerous in the grass family, but are now being discovered in other families as well. For more on C_4-C_3 comparisons, see Black (1971).

As might be expected, C_4 plants dominate desert and grassland communities in warm temperate and tropical climates, and are rare in forests or in the cloudy north where light intensities and temperatures are low. Despite their lower photosynthetic "efficiency" C_3 plants account for most of the world's primary production, presumably because they are more competitive in communities of mixed species where light, temperature, and so on, are average rather than maximum. This may be another example of the principle already mentioned: Survival of the fittest in the real world does not always go to the species that is physiologically superior in monoculture.

Plants that man now depends on for food, such as wheat, rice, and potatoes are mostly C_3 types, which is not surprising since fuel-powered agriculture is a product of northern countries. Crops of tropical origin such as corn, sorghum, and sugar cane are C_4 plants. Agricultural scientists may be missing the boat in not seeking to domesticate additional C_4 plants for use in irrigated deserts and the tropics.

PATTERN OF WORLD DISTRIBUTION OF PRIMARY PRODUCTION

The world distribution of primary production is shown schematically in Figure 3-7. Values represent the average gross production rate per square meter of area to be expected in an annual cycle. As previously indicated, as much as 90 percent of gross production may be available to heterotrophs, but usually only about 50 percent is actually utilized. It should be remembered that man or any other single species cannot assimilate all of the net production. For example, cornstalks and wheat stubble and roots would be included in the total production of these crops, but only the grain is currently consumed by man.

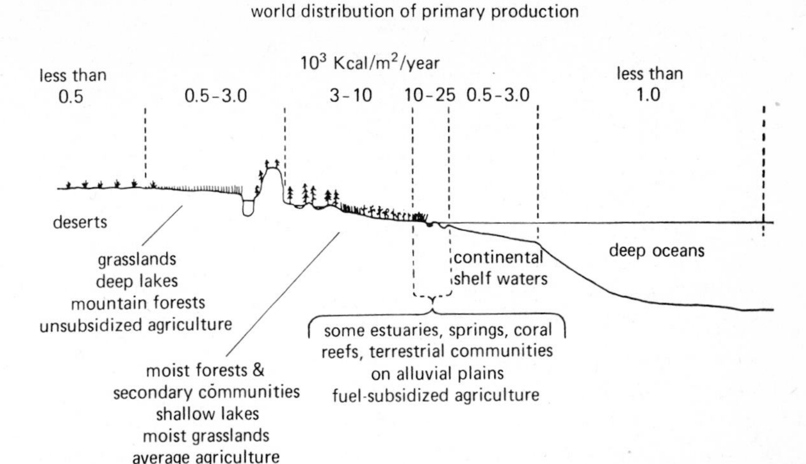

Fig. 3-7 The world distribution of primary production in terms of annual gross production in thousands (10^3) of kcal/m² in major ecosystems. Only a relatively small part of the biosphere is naturally fertile. (Redrawn from E. P. Odum, *Fundamentals of Ecology*, 3rd ed. W. B. Saunders, 1971).

As may be seen from Figure 3-7, there are about three orders of magnitude in potential biological fertility of the world: (1) large parts of the open oceans and land deserts ranging around 1000 kcal/m²/year or less; these are the solar-powered ecosystems that are nutrient- or water-limited. (2) Many grasslands, coastal seas, shallow lakes, and ordinary agriculture range between 1000 and 10,000; these are the energy-subsidized solar-powered systems (compare with Table 2-1). (3) Certain shallow water systems such as estuaries, coral reefs, and mineral springs together with moist forests, intensive agriculture (such as year-round culture of sugar cane or cropping on irrigated deserts), and natural communities on alluvial plains may range from 10,000 to 20,000. Production rates higher than 20,000 have been reported for experimental crops, polluted waters, and limited natural communities. A probable upper limit of 40,000-50,000 has already been noted.

Two tentative generalizations may be made from the data at hand. First, basic primary productivity is not necessarily a function of the kind of producer organism or the kind of medium (whether air, fresh water, or salt water), but is controlled by local supply of raw material, sun energy, supplemental energy, and the ability of local communities as a whole (and including man) to utilize and regenerate materials for continuous reuse. Terrestrial systems are not inherently different from aquatic situations if light, water, and nutrient conditions are similar. However, large bodies of water are at a disadvantage because a large portion of light energy may be absorbed by the water before it reaches the site of maximum mineral supply in the deep water. Secondly, a very large portion of the earth's surface is open ocean or arid and semiarid land and thus in the low production category, because of lack of nutrients in the former and lack of water in the latter. Many deserts can be irrigated successfully, and it is theoretically possible and perhaps feasible in the future to bring up "lost" nutrients from the bottom of the sea and thus greatly increase production at, of course, the expenditure of some form of energy. Such an "upwelling" occurs naturally in some coastal areas, and these have a productivity many times that of the average ocean. A famous example of the effect of upwelling on productivity is found along the coast of Peru. Currents are such that nutrient-rich bottom waters are constantly being brought to the surface so that phytoplankton does not suffer the usual nutrient limitations of the sea. The area supports very large populations of fish and fish-eating birds; so much guano is produced by the birds as they nest along shore that man is able to harvest it for fertilizer on a continuous-yied basis. Ryther (1969) has called the Peruvian upwelling area the world's most productive natural fishery. Some 10^7 metric tons of anchovies are harvested annually from $60 \times \cdot 10^3$ km^2 which comes to about 300 kcal/m^2, a very high *secondary* production. Because the fleets of all fishing nations fish this area, even this bonanza is now in danger of being overfished.

The world distribution of primary production is displayed in more detail in Table 3-1, which also includes an estimate of the global area occupied by major ecosystems, and also an estimate of total gross production of the biosphere. The word "estimate" should be emphasized since there is yet no accurate inventory of productivity on a global basis, although the beginning of such an inventory is underway as part of an "International Biological Program" which is being funded by governmental agencies of many of the nations of the world. For the most recent survey see the symposium entitled "The Primary Production of the Biosphere," edited by Whittaker and Likens (1973).

When the first estimates of global productivity were made in the 1940s it was assumed that the productivity of the ocean was greater than that of the land because it was larger. Then it was discovered that much of the ocean was "desert," so the consensus now is that the land areas contribute more than half of the total. The estimate of 10^{18} kcal/year for global productivity is less than 1 percent of the solar energy entering the biosphere, as we have already noted. But this does not mean that it will be easy to increase world productivity or divert a larger share to food for man, as we shall see in the next section.

Table 3-1. Estimated Gross Primary Production (annual basis) of the Biosphere and Its Distribution Among Major Ecosystems[a]

Ecosystem	*Area* 10^6 *km*^{2}	*Gross Primary Productivity kcal m*^{2}*/yr*	*Total Gross Production* 10^{16} *kcal/yr*
MARINE			
open ocean	326.0	1000	32.6
coastal zones	34.0	2000	6.8
upwelling zones	0.4	6000	0.2
estuaries and reefs	2.0	20,000	4.0
Subtotal	362.4	—	43.6
TERRESTRIAL			
deserts and tundras	40.0	200	0.8
grasslands and pastures	42.0	2500	10.5
dry forests	9.4	2500	2.4
boreal coniferous forests	10.0	3000	3.0
cultivated lands with little or no energy subsidy	10.0	3000	3.0
moist temperate forests	4.9	8000	3.9
fuel subsidized (mechanized) agriculture	4.0	12,000	4.8
wet tropical and subtropical (broadleaved evergreen) forests	14.7	20,000	29.0
Subtotal	135.0	—	57.0
Total for biosphere (round figures) (not including ice caps)	500.0	2000	100.0

[a] Reproduced from E. P. Odum, *Fundamentals of Ecology*, 3rd ed. Saunders, Philadelphia, 1971.

FOOD FOR MAN Approximate yields of major food crops at three levels of auxiliary energy, and three levels of protein content are shown in Table 3-2. Millions of people are not getting enough calories of food; millions of others are not getting enough protein, especially children who have a higher requirement for growth. Thus, it is important to consider both quantity and protein quality. From Table 3-2 we see that world average yield is two to three times less than that of high yields in affluent countries because most of the world's cultivated acres do not have the benefit of high-energy subsidy. Average yields are, in fact, very close to the bottom; that is, world

Table 3-2. Annual Yields of Edible food and Estimated Net Primary Production of Major Food Crops at Three Levels of Energy Subsidy[a] and Three Levels of Protein Content

	Edible Portion		
	Harvest Weight (kg/ha)	*Calorie Content (kcal/m²)*	*Estimated Net Production[b] (kcal/m²)*
A. Sugar cane; low protein content			
1. Hawaii	11,000	4000	12,000
2. Cuba	3300	1000	3700
3. world average	3000	1000	3500
B. Wheat and rice; moderate protein content			
1. Japan and Netherlands	5000	2000	5500
2. India	1500	600	1600
3. world average	2000	800	2400
C. Soybeans; high protein content			
1. Canada	2000	800	2400
2. Indonesia	650	260	800
3. world average	1200	480	1400

[a] 1. Country with high yield (fuel-subsidized agriculture).
2. Country with lowest yield (undeveloped country with little energy subsidy).
3. World average.

[b] Roots and straw (stalks and leaves), plus harvested foods.

Figures in this table are rounded-off averages from *United Nations Production Yearbook* 1970.

averages are little better than that reported from the very poorest countries. To double food yield in the latter requires a tenfold increase in fuel, fertilizers, and pesticides (see Figure 15-2 in E. P. Odum, 1971). Those who think that we can upgrade agricultural production in "undeveloped" countries simply by sending seeds of new varieties and a few "agricultural advisors" are tragically naive. Crops highly selected for industrial agriculture must be accompanied by many calories of fuel, which, of course, the developed countries are hoarding for themselves. A second point to note from Table 3-2 is that, as always, quality comes at the expense of quantity. Yields of a high protein crop, such as soybeans, average one half to one third less than that of low or moderate protein content crops. Getting one's protein from meat, however desirable from the nutritional standpoint, does not help the per-unit-area-yield problem since, as we have seen in our discussion of food chain dynamics, there is at least an 80 percent loss in transfer from grain to meat.

We see from Table 3-3 that in 1970 man harvested about 5.3×10^{15} kcal of food, 99 percent from the land and 78 percent from plants. This

Table 3-3. **Food Harvested by Man, Total for the Biosphere, 10^{12} kcal/year—as of 1967[a]**

	Ocean	*Land*	*Total*
Plant	0.06	4200	4200.06
Animal	59.20	1094	1153.20
Totals	59.26	5294	5353.26

[a] Reproduced from E. P. Odum, *Fundamentals of Ecology,* 3rd ed. Saunders, Philadelphia, 1971.

would theoretically give the approximately 4 billion (4×10^9) people in the world their minimum annual requirement of 1 million (10^6) kcal (see page 15) even allowing for unavoidable waste, if food were evenly distributed. But, of course, it is not, and probably can never be because of the problems of distribution, transportation, economics, and so on.

The estimate of 5×10^{15} kcal harvested is about 1 percent of global net primary production and 0.5 percent of gross (as estimated in Table 3-1). It would seem that man is not yet making much of a dent in the photosynthetic capacity of the earth, but the real impact looks quite different when we consider the following:

1. The oceans are no bonanza since only animal food can be conveniently harvested, and only very limited areas are rich enough to support intensive fisheries. Most fishery experts believe that man is already harvesting all he can get from the natural production of the sea. To "cultivate" the sea on any large scale would require huge investments of energy, the cost-benefits for which might be negative.
2. Since domestic animals outnumber people 5 to 1 in terms of equivalent food requirements, then man plus his animals are taking just about 6 percent of net production of the biosphere or at least 12 percent of that of the land area, and this includes not only cultivated land and pastures, but much "wildland" on which animals graze.
3. All of the best land suitable by natural fertility and slope for intensive "row crop" agriculture is now in cultivation; this amounts to about 12 percent of land area. Only 24 percent of land is truly arable, and to get the additional 12 percent into intensive food production would require energy subsidies much greater than that required for the good land. An additional 25 percent of land is in pastures, much of which are marginal in productivity.

4. To try to cultivate the huge areas of steep land and remot grasslands and deserts is to invite trouble with other neces sary resources, such as water. Attempts to cultivate too mucl land contributed to the failure of past civilizations; even i modern technology is capable of reducing environmenta degradation, the cost remains formidable. The old cliche tha "man does not live by bread alone" must certainly be heeded

Perhaps the best way to view the food problem is to consider i from the per capita viewpoint. To provide the diet now consumed b an American, about 2.5 acres (1 hectare) are required when we con sider land area required to produce meat, orange juice, and leaf vegetables along with staple grains. Another acre is required to pro duce fibers (wood, paper, cotton, and so on). As of 1970 there wer only 10 acres of land per person average for the whole world. I population doubles in the next 50 years there will be only be 5 acre (2 hectares) per capita to provide *all* requirements—water, oxyger waste treatment, fibers, living space, recreation, as well as food.

We cannot hope to do justice to the subject of "food for man in this brief introduction. Relationships are extremely complex an there is much controversy. For further reading we recommend th three-volume treatise *The World Food Problem* and George Borg strom's books (1967, 1969). But we must warn you that these are no easy reading, and that there are no easy "quick-fix" solutions.

The whole gambit of natural and cultivated net primary pro duction is summarized in Table 3-4. When we look at it from th basic energy standpoint there is no difference between man's crop and nature's crops. Given sunlight, nutrients, water, and adapte plants, net production is a function of available supplemental energ —tides in case of salt marsh, and fuel in case of agriculture. Jus because man does not harvest directly the net production of th marsh grass does not mean it is valueless to him. Useful work o waste assimilation and recycling worth many dollars is accomplishe (see page 216) and the seafood flowing off the end of the foo chain (see Figure 3-4) is free except for the cost of harvesting an processing.

NET AND GROSS ENERGY If there is one principle we hope you wi remember and pass on to every person yo meet, it is this: *To a consumer, such as man, only net energy counts* and this applies to *all* energy, food, and fuel. Gross energy, as a poten

Table 3-4. **Annual Net Primary Production of Natural Stands of Marsh Grass and Cultivated Stands of Corn, Both C_4 Plants, as a Function of the Level of Energy Subsidization**

	kcal/m²	*Tons Dry Matter/Acre*
Spartina alterniflora, marsh grass[a]		
streamside stands, vigorous, twice daily tidal irrigation	16,000	17.8
low marsh stands, gentle, frequent tidal irrigation	9200	10.2
high marsh stands, infrequent tidal irrigation (spring tides only)	3000	3.3
Zea Mays, corn[b]		
average, Iowa; heavy subsidy (fuel, fertilizers, pesticides, etc.)	15,000	16.7
average, United States; moderate subsidy	5000	5.6
average, India; low subsidy	2000	2.2

[a] Data from E. P. Odum (1974).
[b] Yield edible grain times 3 as an estimate of total dry matter produced.

tial, is often very impressive in quantity, but must always be evaluated in terms of the amount that can be converted into the desired work. And thermodynamic costs must be less than the net energy obtained if the conversion is to be a long-term benefit. We have seen that the gross energy of solar radiation is huge, but the net energy of food is very small. Likewise, "proven reserves" of oil and coal ("enough to last many centuries" you may be told) are gross energy. The relevant question is how much will be left to power your car and run your city and what will be the price? If mining, extracting, shipping, and processing oil from under the sea, or that locked in shale rock, requires more energy than the final product is worth, there may be no net energy. Likewise for agriculture; already the energy value of some crops is less than the fuel energy required to produce them. Likewise with atomic energy. The potential energy in the atom is fantastic, but so are the costs of converting it. As with most scientists I was once more enthusiastic than I am today about atomic energy replacing fossil fuels within this century. Costs and difficulties have been greater than our best minds predicted, and we may have invested in the wrong kind of atomic energy, because of our preoccupation with atomic bombs (for more on the kinds of atomic energy see page 207). We must continue a massive research effort but it is going to take more time to prove out various possibilities for harnessing atomic energy on the scale we now extract energy from fossil fuels. Which means that we should "power down" and be more effi cient in the use of "proven net energy," at least for a while.

To put some of this in perspective let us consider the fuel-energ budget of the United States as of the early 1970s. From the larg world reserves of fossil fuel the United States receives about 16×10^{1} kcal/year, but only about half (50 percent) is actually converted int useful work. Thus, the "net" is 8, not 16, $\times 10^{15}$, not 10^{x} *quantities* i the ground. Cost of extracting and processing goes up as gross supplie dwindle and we have to turn to lower quality materials. Two of ou most convenient and worthwhile uses of energy are automobiles an electricity, but both are only about 30 percent efficient which con tributes to the low overall efficiency. It seems likely that the 1970s wil see us give up some convenience to improve efficiency and thus stretc out fuel supplies as they become more expensive to convert from gros reserves. We have already spoken of need for diversification in situa tions such as this (page 56).

There is a parallel of sorts in economics. For many years th Gross National Product (GNP) has been considered a good index o economic well-being. Now, many economists are suggesting that Ne Economic Worth (NEW) would be a much better measure. In com puting NEW the "bads" (pollution costs, and so on), as well as th "goods" (manufactured products, and so on) are considered, an maintenance work, such as the work of the housewife, is included. I recent years the GNP of most nations has been going up but the NEV for the United States has leveled off, indicating that the real economi situation has not been improved by ever larger production of. har goods.

Again, these comments are only suggestive. The situation is com plex with many considerations other than strictly economic ones. Fo further reading see Nordhaus and Tobin (1972); Daley (1972); an Georgescu-Roegen (1972).

SUGGESTED READINGS

References cited

Black, C. C. 1971. Ecological implications of dividing plants into group with distinct photosynthetic production capacities. *Adv. Ecol. Res* (J. B. Cragg, ed.), 7:87–114.

Borgstrom, George (see "food for man" list below)

Bray, J. R. 1961. Measurement of leaf utilization as an index of minimun level of primary consumption. *Oikos*. 12:70–74.

Daly, Herman E. 1972. *Towards a Steady-State Economy. Introduction*, pp. 1–29. San Francisco: W. H. Freeman. (See also in *Patient Earth*, ed. Harte and Socolow, 1971. New York: Holt, Rinehart and Winston.)

Gates, David M. 1963. The energy environment in which we live. *Amer. Sci.* 51:327–348.

Forrester, Jay W. 1971. *World Dynamics.* Cambridge, Massachusetts: Wright-Allen Press.

Georgescu-Roegen, N. 1972. The entropy law and the economic problem. In *Towards a Steady-State Economy*, ed. H. E. Daly, pp. 37–49. San Francisco: W. H. Freeman. (Also full length book, Cambridge: Harvard Univ. Press, 1971.)

Nordhaus, William and James Tobin. 1972. Economic growth: is growth obsolete? Colloquium V, Nat. Bureau Economic Research. Columbia Univ. Press. (Suggests that NEW, National Economic Welfare, which considers economic "bads" as well as "goods" is a better index than GNP.)

Odum, Eugene P. 1971. *Fundamentals of Ecology*, 3rd ed. Philadelphia: Saunders.

________1974. Halophytes, energetics and ecosystems. In *Ecology of Halophytes,* eds. Reimold and Queen New York: Academic Press.

Odum, Eugene P. and Armando A. de la Cruz. 1967. Particulate organic detritus in a Georgia salt marsh-estuarine ecosystem. In *Estuaries*, G. Lauff ed., *Amer. Assoc. Adv. Sci.* 83: 383–388.

Odum, H. T. 1971. *Environment, Power and Society.* New York: John Wiley & Sons.

Odum, H. T.; W. M. McConnell; and W. Abbott. 1958. The chlorophyll "A" of communities. *Publ. Inst. Marine Sci., Univ. of Texas.* 5:65–97.

Odum, H. T. and R. C. Pinkerton. 1955. Times speed regulator, the optimum efficiency for maximum output in physical and biological systems. *Amer. Sci.* 43:331–343.

Odum, William E. and Eric J. Heald. 1972. Trophic analysis of an estuarine mangrove community. *Bull. Marine Sci.* 22(3):671–738.

Riley, Gordon A. 1956. Production and utilization of organic matter in Long Island Sound. *Bull. Bingham Oceanogr. Coll.* 15:324–344.

Ryther, John H. 1969. Photosynthesis and fish production in the sea. *Science.* 166:72–76.

Teal, J. M. 1962. Energy flow in the salt marsh ecosystem of Georgia. *Ecol.* 43:614–624.

Whittaker, Robert H. and G. E. Likens, eds. 1973. Primary production of the biosphere. *Human Ecol.* 1:301–369.

World Food Problem (see "food for man" list below)

Thermodynamics

Morowitz, H. J. 1970. *Entropy for Biologists. An Introduction to Thermodynamics.* New York: Academic Press. (Intended to be understandable for the undergraduate student.)

Schrodinger, Erwin. 1967. *What is Life*? Chapter 6: Order, disorder and entropy, pp. 72–80. Cambridge, England: Cambridge Univ. Press.

Productivity and food chains

Phillipson, John. 1966. *Ecological Energetics*. New York: St. Martin's Press. (Elementary introduction for college students.)

Odum, Eugene P. 1962. Relationships between structure and function in the ecosystem. *Jap J. Ecol.* 12: 108–118. (Reprinted in G. W. Cox, ed. *Readings in Conservation Ecology*, 2nd ed. 1974.) New York: Appleton-Century-Crofts.

________. 1968. Energy flow in ecosystems: a historical review. *Amer. Zool* 8:11–18.

Whittaker, R. H. 1970. *Communities and Ecosystems*. New York: Macmillan. (See also Whittaker and Likens, references cited above.)

Wiegert, Richard G. 1970. Energy transfer in ecological systems. A Biological Sciences Curriculum Study pamphlet. Chicago: Rand-McNally (Introduction for high school students.)

Photosynthesis

Black, C. C. (See "references cited" above.)

Bjorkman, Olle and Joseph Beery. 1973. High-efficiency photosynthesis. *Sci Amer.* 229(4):80–93.

Energy for man

Cook, Earl. 1971. The flow of energy in an industrial society. *Sci. Amer* 224(3):135–144.

Energy. Special issue of *Sci. Amer.* Vol 224, Sept. 1971. Also special issue of *Science*. Vol. 184, No. 4134, April 19, 1974.

Odum, Howard T. 1973. Energy, ecology and economics. *AMBIO* (Royal Swedish Academy of Sciences) 2(6):220–227. (Reprinted and available on request from: Environmental Information Center, Florida Conservation Foundation, Winter Park, Fla. Concept "net energy" in terms of amount and cost of the energy required to produce energy for human use. See also: *Business Week*. June 8, 1974, pp. 88–89.)

Singer, S. F. 1970. Human energy production as a process in the biosphere *Sci. Amer.* 223(3):175–190.

Weinberg, Alvin and R. P. Hammond. 1970. Limits to the use of energy *Amer. Sci.* 58:412–418. (Suggest that social, economic, and environmental considerations, not technology, may limit man's use of atomic energy.)

Food for man

Borgstrom, George. 1967. *The Hungry Planet.* New York: Collier. 1969. *Too Many.* New York: Macmillan.

Brown, Lester R. 1970. Seeds of Change: the Green Revolution and Development in the 1970's. New York: Praeger. (Analysis of impact of new genetic varieties of food crops on development strategy of nations.) See also, *Sci. Amer.* 223(3):160–170.

Emory, K. D. and C. O. D. Iselm. 1967. Human food from ocean and land. *Science.* 157:1279–1281.

Idyll, C. P. 1970. *The Sea Against Hunger.* New York: Thomas Crowell. (Well-illustrated inventory of food resources of sea and future potential of marine culture.)

Perelman, M. J. 1972. Farming with petroleum. *Environment.* 14:8–13.

Pimentel, D.; L. E. Hurd; A. C. Bellotti; M. J. Forster; I. N. Oka; O. D. Sholes; and R. J. Whitman, 1973. Food production and the energy crisis. *Science.* 182:443–149.

(These two papers document the dependence of modern agriculture on fuel energy subsidies; see also Fig. 8-2, this book.)

"The World Food Problem," a Report of the President's Science Advisory Committee Panel on world food supply. The White House, Washington, 3 Volumes, 1967. (The best available comprehensive and objective review.)

chapter 4

Biogeochemical Cycles and Limiting Factor Concepts

In the preceding chapter important principles and some orders of magnitude regarding energy flow within ecosystems were discussed. As already emphasized, the movement of materials in the ecosystem is an equally important consideration. The more or less circular paths of the chemical elements passing back and forth between organisms and environment are known as *biogeochemical cycles*. "Bio" refers to living organisms and "geo" to the rocks, soil, air, and water of the earth. Geochemistry is an important physical science, concerned with the chemical composition of the earth and the exchange of elements between different parts of the earth's crust and its oceans, rivers, and so on. Biogeochemistry is thus the study of the exchange (that is, back and forth movement) of chemical materials between living and nonliving components of the biosphere.

In Figure 4-1 a biogeochemical cycle is superimposed on a sim-

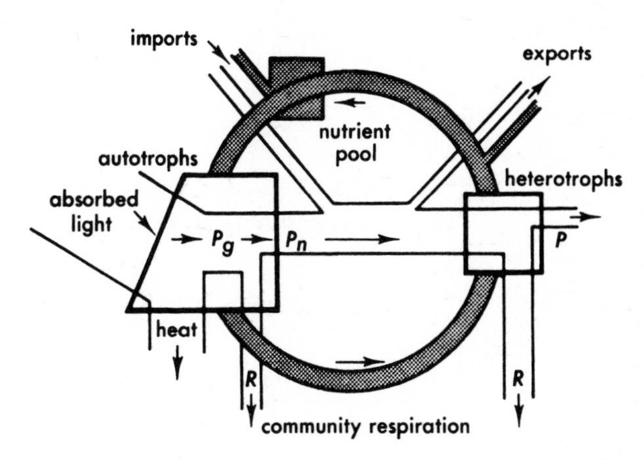

Fig. 4-1 A biogeochemical cycle (stippled circle) superimposed upon a simplified energy-flow diagram, contrasting the cycling of material with the one-way flow of energy.

plified energy-flow diagram to show the interrelation of the two basic processes, and to reemphasize the point already made, namely that energy is required to drive the cycling of materials. Vital elements in nature are never, or almost never, homogeneously distributed or present in the same chemical form throughout an ecoystem. Rather, materials exist in compartments or pools, with varying rates of exchange between them. From the ecological standpoint it is advantageous to distinguish between a large, slow-moving nonbiological pool and a smaller but more active pool that is exchanging rapidly with organisms. In Figure 4-1 the large reservoir is the box labeled "pool"; and the rapidly cycling material is represented by the stippled circle going from autotrophs to heterotrophs and back again. Sometimes the reservoir portion is called the unavailable pool and the cycling portion the available pool; such a designation is permissable provided it is clearly understood that the terms are relative. An atom in the reservoir pool is not necessarily permanently unavailable to organisms but only relatively so; in comparison, an atom in the cycling pool is instantly available. Almost always there is a slow movement of atoms between the unavailable and the available pools.

Decomposition not only releases nutrients but the organic byproducts may also increase the availability of minerals for uptake by autotrophs. One way this occurs is by a process known as *chelation* (fr. chele = claw, referring to grasping) in which organic molecules "grasp" or form complexes with, calcium, magnesium, iron, copper, zinc, and other ions. Such chelated minerals are more soluble and less toxic than some of the inorganic salts of the element, especially in the case of metals. Chelators are nearly always added to cultures and microcosms to enhance the availability of nutrients.

AMOUNTS VERSUS RATES When a farmer sends a sample of soil to the soils laboratory of his state university for routine testing, the sample is often treated with 0.1 normal acid or alkali solution. The quantity of minerals, such as phosphorus, calcium, or potassium, removed by such gentle treatments is considered a crude measure of quantities available to plants (that is, the size of the available pools). A simple test such as this may provide a useful basis for fertilizer recommendations, but as often as not leaves much to be desired. As with energy, it is evident that the rates of movement or cycling may be more important in determining biological productivity than the amount present in any one place at any one time. Turnover time and turnover rate, as convenient measures of flux, have already been discussed (see page 28). It is the *flux* (= rate of transfer), rather than the concentration, that is of prime importance. Tracers have been a great help in determining rates of movements since tagged atoms can actually be followed as they exchange with organism and environment.

During the past ten years what has come to be known as *mineral cycling* has received increasing attention in ecological research. One-at-a-time study of single elements has been superseded by studies of the behavior of groups of linked elements and compounds as related to energy flow and stability. These newer approaches have been made possible by improved techniques of systems modelling, as described in Chapter 1. Major breakthroughs have been made in our understanding of the role of energy as the driving force for cycling, and the role played by microorganisms in releases of essential elements from otherwise unavailable pools. For a review of these general considerations, see Pomeroy (1970).

NUTRIENTS, NONESSENTIAL ELEMENTS, AND POISONS As already noted in Chapter 2, elements and dissolved salts essential to life may be conveniently termed *biogenic salts* or *nutrients* and divided into two groups, the *macro-*

nutrients and the *micronutrients.* The former include elements and their compounds that have key roles in protoplasm and that are needed in relatively large quantities, as for example, carbon, hydrogen, oxygen, nitrogen, potassium, calcium, magnesium, sulfur, and phosphorus. The micronutrients include those elements and their compounds also necessary for the operation of living systems but are required only in very minute quantities; for example, iron, manganese, copper, zinc, boron, sodium, molybdenum, chlorine, vanadium, and cobalt. It should be emphasized at this point that elements that have no known biological function also circulate between organisms and environment. These may enter biogeochemical cycles linked with essential elements by reason of chemical affinity, or they may be simply carried along in the general energy-driven stream. Likewise, poisons produced by man, such as insecticides and radioactive strontium, all too often enter vital cycles and become lodged in the tissues of animals and man (more about this later). Although organisms do develop adaptive mechanisms to exclude harmful substances, there is no way that living membranes can function efficiently in the exchange of vital materials, and, at the same time be completely selective as to what is "good" and what is "bad." Even if harmful substances are not lethal, an energy stress is placed on the organism since it must expend extra energy to sequester or "pump out" the poison.

Nonessential elements are, therefore, of great ecological importance if they occur in quantities or forms that are toxic, if they react to bind or make unavailable essential elements, or if they are radioactive. Thus, the ecologist is concerned with nearly all of the natural elements of the periodic table as well as with the newer man-made ones, such as plutonium.

THE SULFUR CYCLE IN AQUATIC ENVIRONMENTS

The sulfur cycle as diagrammed in Figure 4-2 illustrates the main features of a biogeochemical cycle in a specific ecosystem. The cycling and reservoir pools, the chemical forms of the elements, and the organisms involved are all shown in the diagram. Sulfate (SO_4) in the water is the principal available form that is reduced by autotrophic plants and incorporated into proteins, sulfur being an essential constituent of certain amino acids (path 1, in Figure 4-2). When animals excrete, or the bodies of plants and animals are decomposed by heterotrophic microorganisms, sulfate may be returned to the water (path 3) or hydrogen sulfide (H_2S) is released (path 2). Some of the H_2S is then reconverted to sulfate by specialized sulfur bacteria (paths 4, 5, 7). Some of these bacteria are called *chemosynthetic*

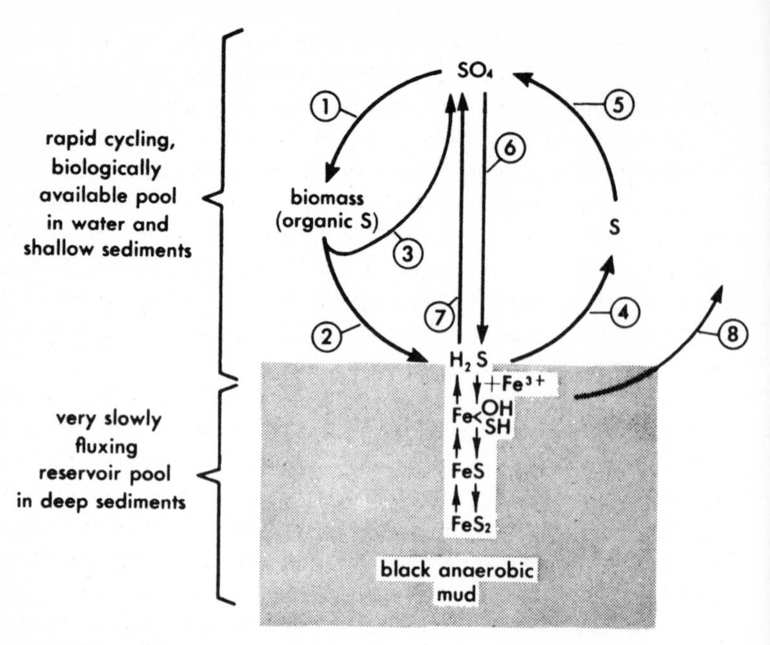

Fig. 4-2 The sulfur cycle in an aquatic system. Organisms play key roles in the rapidly cycling pool as follows: (1) primary production by autotrophs; (2) decomposition by heterotrophic microorganisms; (3) animal excretion; (4), (5) steps by specialized colorless, purple, and green sulfur bacteria; (6) desulfovibrio bacteria (anaerobic sulfate reducers); (7) thiobacilli bacteria (aerobic sulfide oxidizers). Step 8 releases phosphorus (from insoluble ferric phosphate), thus speeding up the cycling of this vital element.

organisms, because they obtain their own energy from the chemical oxidation of inorganic compounds (in this case oxidation of sulfide to sulfur, and so on) instead of from light as do photosynthetic organisms or from organic matter as do heterotrophic organisms. The green sulfur bacteria are photosynthetic, but since H_2S is oxidized rather than H_2O, as in regular photosynthesis, free oxygen is not released. An equation for this type of bacterial photosynthesis is obtained by substituting H_2S for H_2O on the left side, and 2S for O_2 on the right side of the equation on page 76 . Under anaerobic conditions much of the H_2S may not be oxidized but may pass into the reservoir pool where back and forth fluxing is very slow, as shown in Figure 4-2. Interestingly, when iron sulfide compounds are formed, phosphorus is converted from insoluble to soluble form and thus becomes available to living organisms in the water. Here is an excellent illustration of how one nutrient cycle regulates another. The sulfur cycle in terrestrial environments is much the same as in aquatic ones—the soil being the site of major fluxes, as shown in Figure 4-2.

That organisms, especially microorganisms, play key roles is true not only in the sulfur cycle but in the nitrogen cycle and most of the others. Organisms are not just passive actors in a physical and chemical milieu, but are active participants in the regulation of their own environment. No one organism or population alone has much control, of course, but the sum total of processes in the well-ordered ecosystem ensures continuous supplies of materials and energy needed for life. Redfield (1958), for example, has marshalled the evidence to indicate that organisms over long periods of time have largely controlled the chemical composition of the sea, likewise, for the atmosphere. The earth's original atmosphere, as derived from purely geological processes, contained little or no oxygen and much carbon dioxide until green plants evolved and began releasing the former and removing the latter (see Berkner and Marshall, 1966, for an account of the evolution of the atmosphere). As a dependent heterotroph man cannot alone control the biosphere for his own good; he must have the cooperation of the "germs" of the soil and water as well as the autotrophs and many other organisms. As we have seen, a diversity of at least five types of bacteria are required to recycle the vital element, sulfur. Although bacteria and other unseen organisms that work for the good of man and nature are adaptable and tough, they are composed of living protoplasm which is as vulnerable to poisons as man himself. Too often, man works to obtain a temporary advantage by increasing the rate of flow of materials, but forgets to arrange for the return mechanism, and through ignorance and carelessness hampers the natural return mechanisms.

TWO BASIC TYPES OF BIOGEOCHEMICAL CYCLES

From the standpoint of the biosphere as a whole, biogeochemical cycles fall into two groups: the sedimentary types as illustrated by the sulfur cycle and the gaseous type as illustrated by the nitrogen cycle, shown in Figure 4-3. In the latter the air, rather than the soil and sediments, is the great reservoir and safety valve of the system. Nitrogen is continually feeding into and out of this great reservoir from the rapidly cycling pool associated with organisms. Both biological and nonbiological mechanisms are involved in denitrification and nitrogen fixing, the latter being the conversion of the nitrogen of the air, which is not available to autotrophs, to nitrates, which are. As in the case of the sulfur cycle, specialized microorganisms play key roles. For example, only a relatively few species of bacteria and blue-green algae, which, fortunately, are very abundant in many systems, can fix nitrogen. No so-called higher plant or animal has this

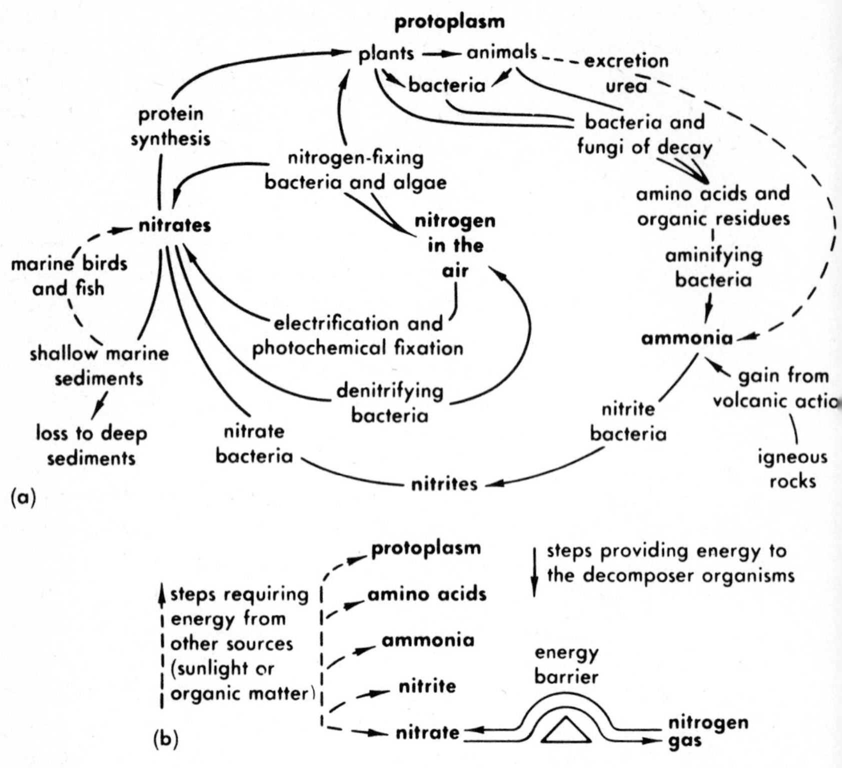

Fig. 4-3 The nitrogen cycle is an example of a biogeochemical cycle with a large gaseous reservoir and many feedback mechanisms. (Redrawn from E. P. Odum, *Fundamentals of Ecology,* 2nd ed. W. B. Saunders, 1959.)

ability. Legumes fix nitrogen only through the specialized bacteria that live in their roots. The self-regulating feedback mechanisms, as shown in a very simplified manner in Figure 4-3, make the nitrogen and other gaseous-type cycles (such as the carbon or water cycles) relatively perfect in terms of large areas of the biosphere. Any increase in movement along one path is quickly compensated for by adjustments along other paths. Locally, however, nitrogen often becomes limiting to the biological system either because regeneration (that is, movement from unavailable to available) is too slow or a net loss is occurring from the local system.

Most nutrients are more earthbound than nitrogen; their cycles are less perfect and, consequently, more easily disrupted by man. The phosphorus cycle is a good example of a sedimentary cycle of the utmost importance. Phosphorus is required for nearly all the basic energy transformations that distinguish living protoplasm from non-living systems, and it is relatively rare on the earth's surface in terms of biological demand. Organisms have evolved many hoarding devices for this element; hence, the concentration of phosphorus in a gram of biomass is usually many times that in a gram of surrounding environment (water, for example). The beautiful and efficient way in which the ATP system works within the cell to conserve both energy and materials is described in Loewy and Siekevitz *Cell Structure and Function*, 2nd ed. As the cycling pool of phosphorus spins around in the organism and in local biogeochemical cycles there is nevertheless a tendency for a slow downhill movement of the reservoir phosphorus, following the pattern of erosion and sedimentation. In the long range, return or replacement occurs as a result of both physical and biological processes. Weathering of rocks, airborne dust, volcanic gases (that is, natural fallout), and bits of salt spray picked up by the wind are routes that may move small amounts of material uphill. Upwelling of deep ocean waters bringing phosphorus from the unlighted depths to the photosynthetic zone is a very important return mechanism in the sea. The fish-eating guano birds, mentioned in Chapter 3, that annually excrete many tons of phosphorus on their nesting grounds on the west coast of South America are examples of agents in biological recycling. Hutchinson (1948) has shown that return of phosphorus from the sea to land as the result of fish harvested by birds and man is of no small magnitude. In well-ordered systems such as a coral reef (see also Chapter 2) one is impressed with the numerous biological mechanisms that keep the net loss of materials to a very minimum amount, perhaps no more than is returned by natural means. In many areas, however, man has so increased the rate of erosion that the one-way movement of phosphorus into the large unavailable pool in the deep ocean sediments has been increased. Hutchinson estimates that at the present time natural means of return are inadequate to keep up with the downhill loss. For the present, agricultural man is not particularly worried, since he is able to mine the considerable reserves of phosphate rock and replace some of the loss. Soon, however, it will be necessary to improve the retention and recycling within man-ordered systems or else we will have to work to return materials from the deep sea, which will cost us energy and probably raise the price of food.

AIR POLLUTION AND THE SULFUR AND NITROGEN CYCLES

Both the sulfur and the nitrogen cycles are involved in urban and industrial air pollution that increasingly threatens the health of man and plants. The oxides of nitrogen (NO and NO_2) and sulfur (SO_2) are rather toxic gases which are but transitory steps leading to the formation of nitrate (NO_3) and sulfate (SO_4) in their respective cycles. They are normally present in most environments in very low concentrations. The burning of fossil fuels, however, has greatly increased the concentration of these volatile oxides in the air, especially in urban regions, to the extent that they constitute about a third of the some 150 million tons of industrial air pollutants estimated to be released annually over the United States. Coal-burning electrical generating plants constitute a major source of SO_2, and the automobile is a major source of NO_2. Sulfur dioxide is very damaging to the photosynthetic process; the destruction of vegetation and crops around copper smelters and in the Los Angeles basin is caused by this pollutant. Also, SO_2 combines with water vapor to produce airborne sulfuric acid (H_2SO_4). The resulting "acid rain" causes millions of dollars damage to metal, limestone, and other construction material, not to mention possible damage to respiratory tissues. Likens and Bormann (1974) have documented how the pH of rainwater falling over the industrialized areas has decreased (that is rain has become more acid) during the past two decades.

The oxides of nitrogen are particularly harmful to lungs and are thought to contribute to the increase in respiratory diseases and cancer in urban populations. Furthermore, these oxides combine with hydrocarbons and other pollutants under the driving force of sunlight to produce a synergism (= total effect of the interaction exceeds the sum of the effects of each substance) called "photochemical smog." Technology is available to reduce these air pollutants at their sources, but so far it has proved to be expensive. Air pollution may well limit the size of fuel-powered cities, at least for the next couple of decades, and this could work out as a net benefit for man.

NUTRIENT CYCLING AND WATERSHEDS

Recent experimental studies on discrete watersheds have greatly increased our understanding of nutrient cycling. For our purposes a watershed can be defined as a landscape catchment basin, including terrestrial slopes, streams, and lakes, from which all runoff comes out through a common stream outlet. In terms of study and management by man a watershed is a convenient-sized ecosystem with definable boundaries. Especially interesting research is being carried out at the Hubbard Brook Experimental Watersheds in New Hampshire, the Coweeta Watersheds in western North Carolina, and similar research areas in other geographical regions of this continent and elsewhere in the world. In these outdoor laboratories the inputs and outputs of water, nutrients, energy, and so on are carefully measured.

Figure 4-4 is a diagram of the calcium budget for one of the

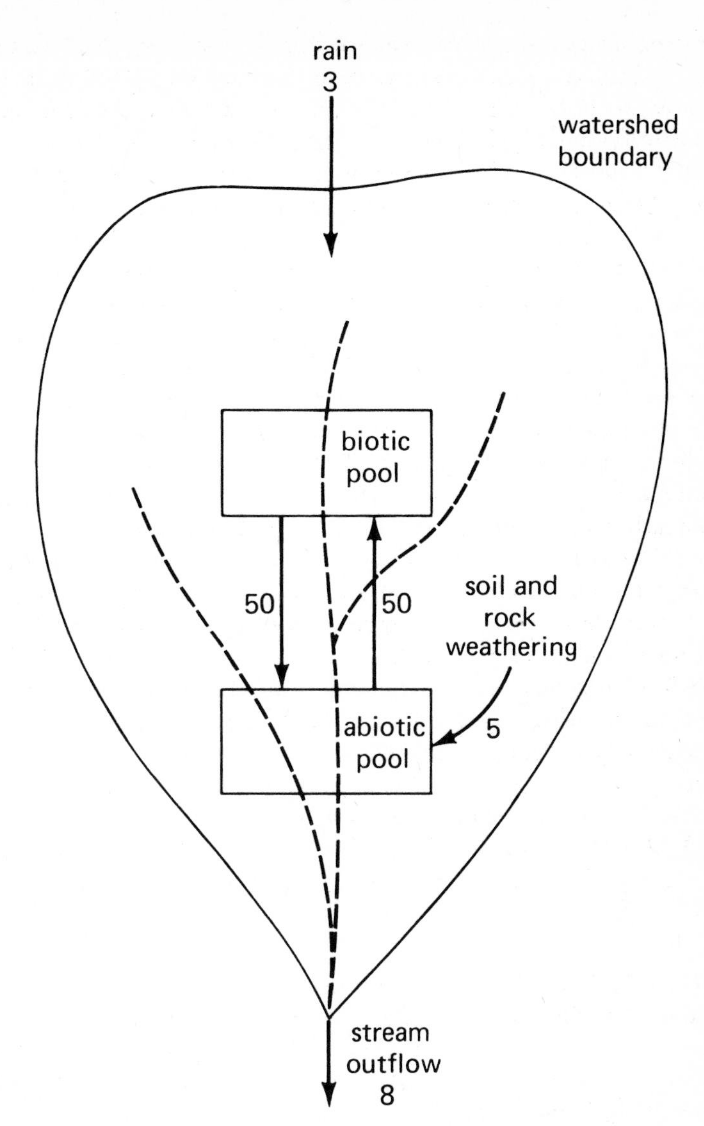

Fig. 4-4 The balanced calcium budget of a forested watershed in New Hampshire (Hubbard Brook Experimental Forest). Figures are calcium flows in kilograms per hectare per year. The inputs and outputs are small as compared to exchanges between biotic and abiotic pools within the ecosystem. (Data of Bormann and Likens, 1967.)

Hubbard Brook watersheds that is covered with an undisturbed forest. Retention by, and recycling within, the undisturbed forest is so effective that the estimated loss from the ecosystem is only 8 kg/hectare/year of calcium (and equally small amounts of other nutrients). Since 3 kg/ha of this is replaced in rain, only an input of 5 kg/ha is needed to achieve a balance, and this is supplied by the normal rate of weathering of the underlying rock that constitutes the "reservoir pool." When the vegetation on one of the experimental watersheds was felled and regrowth the next season was suppressed by herbicides, stream outflow increased and the loss of nutrients was 3 to 16 times (depending on the nutrient) that of adjacent undisturbed watersheds. This large-scale experiment illustrates a common situation where a manipulation that changes the movement of one material (water, in this case) along one pathway also results in changes in other material flows. Increased water yield following removal or reduction of vegetation cover might be advantageous to a city downstream that is short of water, but if such an increased water supply comes at the expense of impoverished productivity of the land (due to unbalanced nutrient budget) and a decrease in water quality downstream (due to increase in mineral and sediment content), then there may be no net advantage to man. Accordingly, for every proposed manipulation we must inquire into the "trade-off" costs. This is, in fact, what is now required by the National Environmental Policy Act (NEPA) recently passed by the U.S. Congress.

For greater details on the Hubbard Brook experiments, see Bormann and Likens (1967); Likens, Bormann, and Johnson (1969); and Likens and Bormann (1972).

GLOBAL CYCLES Now let us move to the global level. Figures 4-5 and 4-6 are models of what are probably the two most important cycles as far as man's impact on the biosphere is concerned, namely the CO_2 (carbon dioxide) cycle and the H_2O (hydrological) cycle. Both are characterized by small, but very active, atmospheric pools that are vulnerable to man-made perturbations which, in turn, can advertently or inadvertently change weather and climates.

Looking at the CO_2 cycle first (Figure 4-5) we see that the carbonate system of the sea and the earth's green belts are very efficient in removing CO_2 from the atmosphere (note the large flux rates labeled "biotic" and "biotic and chemical"). However, the spiraling increase in consumption of fossil fuels coupled with the decrease in the "removal capacity" of the "green belt" is beginning to have an effect on the atmospheric compartment. A basic principle here is that it is the small, active compartment, not the large reservoirs that are first affected by changes in fluxes or "throughputs" (a term encompassing both input and output). Thus, it is the small CO_2 atmospheric

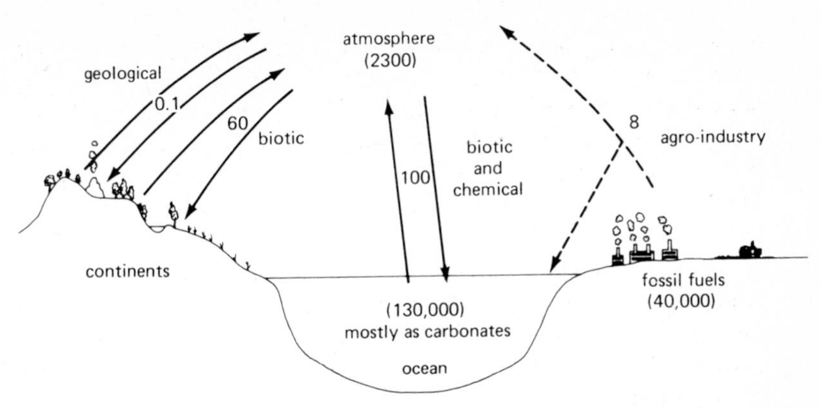

Fig. 4-5 The global CO_2 cycle with estimates of amounts in three major compartments (atmosphere, oceans, and fossil fuels), and flux rates (as shown by arrows) between compartments. Figures of units of 10^9 tons of CO_2. (Data from Plass, 1959; diagram after E. P. Odum, 1971.)

pool, not the much larger oxygen pool, that is of immediate concern.

Strange to say, modern agriculture adds to the CO_2 in the atmosphere since the CO_2 fixed by crops does not compensate for that released from the soil, especially as results from frequent plowing. The rapid oxidation of humus and release of CO_2 normally held in the soil is also having a detrimental effect on the release of micronutrients from underlying rocks because reducing acidity of water percolating deeply in soils reduces the normal weathering of underlying rocks.

Although the rate of release of CO_2 by industry and agriculture is yet fairly small compared with the exchange with the sea (see Figure 4-5), the CO_2 content of the air is slowly rising. Since CO_2 acts like a greenhouse in that it lets light in but retards the flow of heat out of the biosphere, undesirable changes in climates can result, as already noted (page 63). This would be especially likely if the huge reserve of fossil fuel were to be burned in a short time, releasing an estimated 40×10^{12} tons of CO_2. Here is another reason for being frugal and efficient in the use of fossil fuels. What we can apparently expect for the next few decades are new and uncertain, perhaps unstable, balances involving increasing atmospheric CO_2, which heats up the earth, and increasing particulate pollution (dust), which reflects incoming radiation and thus cools the earth. It is extremely important that a network of worldwide measurements be set up to detect changes in CO_2 so as to continually monitor this delicate balance. Technology is available to shift balances if we can be sure which way they are tipping. Unfortunately, international bodies such as the United Nations are so preoccupied with short-term political problems that this common need has yet to receive the attention it deserves.

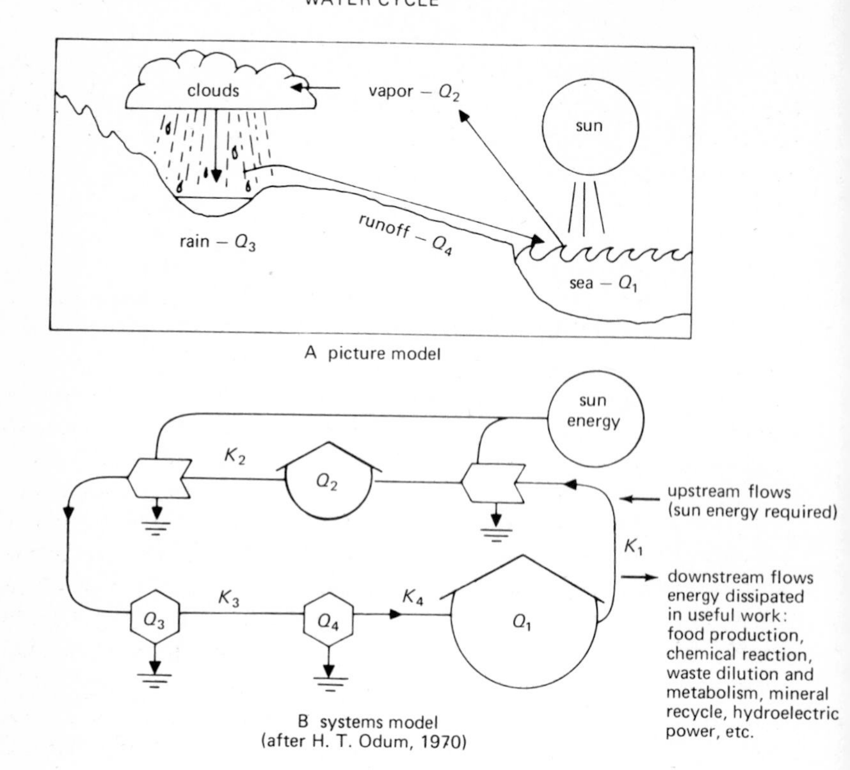

$$Q_1 : Q_2 : Q_3 : Q_4 = \frac{1}{K_1} = \frac{1}{K_2} = \frac{1}{K_3} = \frac{1}{K_4}$$

Fig. 4-6 The hydrological cycle shown as a pictorial model(A); a flow chart model, (B); and a simplified equation model, (C). Storages are represented by Q's and flows by K's. In diagram (B) the storage bin symbols represent the two major water compartments (Q_1, Q_2, while the hexagons (Q_3, Q_4) represent biotic and other components that benefit from downstream flows with appropriate energy losses, as shown by heat-sink symbols. Solar energy drives the upstream flows K_1 and K_2. (After H. T. Odum, 1970.)

The diagrams of the hydrological cycle in Figure 4-6 are set up to emphasize (1) the role of solar energy as the driving force that pumps water uphill, as it were; and (2) the enormously useful work done by water as it runs downhill again. The diagrams also show, again, how modelling proceeds logically from pictures to circuit diagrams to mathematical equations. We have already emphasized that natural recycling of water is "free" and constitutes a vast, but almost unrecognized, use of solar energy (see page 63). It would be sad, indeed, for man to have to use scarce and expensive fuels for this purpose when such fuels are needed to drive the machine of commerce. H. H. Odum (1970) has estimated that a gallon of filtered, clean water delivered to your home would cost two dollars if we had to use fuels to recycle it. With nature doing most of the work your water now costs less than one dollar for a thousand gallons.

BIOLOGICAL MAGNIFICATION While energy decreases and becomes more dispersed at each step in the food chain, some substances become more concentrated with each link in a process that has become known as *biological magnification*. The dramatic increase in concentration of four dangerous substances in four different ecosystems is shown in Table 4-1. In each case the amount in the water or soil (that is, the general environment) at the time of measurement was extremely small and certainly "harmless," but by the time the poisonous substances reached the end of the food chain, concentrations were high enough to cause sickness and death. Physical, chemical, and biological processes may all contribute to the concentration process. DDT may be absorbed through the skin of a fish as well as be ingested with food. It becomes concentrated in vertebrates because it tends to become stored in fat deposits. Radioactive phosphorus is concentrated simply because of the natural tendency of biota to concentrate and store this scarce but vital element, as noted earlier in this chapter. Strontium-90 concentrates in bone, where its radiation may damage sensitive blood-making tissue, because of its chemical resemblance to calcium. In other cases reasons for concentrations are unknown. We must observe, experiment with, and model the ecosystem level because neither laboratory experiments nor general theory are adequate for predicting the behavior of each of the thousands of new chemicals that man is constantly introducing. Concentration factors tend to be higher in aquatic than in terrestrial systems since water is a more "dilute" medium than soil.

Biological magnification has come as a great surprise to physical scientists and technologists who were enthusiastic about the idea that "the solution to pollution is dilution" or, in other words, the belief that poisons would be quickly lost in the vast confines of nature. Because predatory birds are being wiped out by DDT and man himself threatened (since he cannot escape being part of food chains) society has been forced to consider reducing, or banning outright, this pesticide that was once heralded as the solution to all insect pest problems. Likewise, the difficulty of dispersing or containing large amounts of radioactive wastes without contaminating man's food chain

Table 4-1. Concentration Factors (Ratio Amount in Organisms to Amount in Environment) of a Widely Used Persistent Pesticide and Certain Radionuclides Released Near Atomic Reactors

DDT in Long Island Estuary (1967)[a]		*Radioactive Phosphorus (^{32}P) in Columbia River (1956)*[b]		*Radioactive Strontium (^{90}Sr) in a Canadian Lake Receiving Atomic Wastes (1963)*[c]		*Radioactive Iodine (^{131}I) in a Washington Desert Following Stack Release (1956)*[b]	
Water	1	Water	1	Water	1	Vegetation	1
Plankton	800	Insects	3	Sediments	200	Thyroids of jackrabbits	500
Minnows	11,600	Swallows	75,000	Aquatic plants	300		
Predatory fish	34,600	Duck eggs	200,000	Minnows	1000		
Fish-eating birds	92,000			Perch bone	3000		
				Muskrat bone	3900		

[a] Data of Woodwell, Wurster, and Isaacson (1967).
[b] Data of Hanson and Kornberg (1956).
[c] Data of Ophel (1963).

is one major reason why you cannot as yet enjoy the peaceful uses of atomic energy. For more on this important subject see Woodwell (1967).

NUTRIENT CYCLING IN THE TROPICS The pattern of nutrient cycling in the tropics is, in several important ways, different from that in the temperate zone. In the latter a large portion of organic matter and available nutrients is at all times in the soils and sediments; in the tropics a much larger percentage is in the biomass, and is recycled within the organic structure of the system, not in the soil. This contrast is illustrated by the percentage distribution of nitrogen in a temperate and a tropical forest, as shown in Table 4-2.

Table 4-2. Distribution of Cycling Nitrogen in Biotic and Abiotic Pools in a Temperate and a Tropical Forest[a]

	Total Amount g/m²	*Percent in Biomass*	*Percent in Soil*
Temperate Forest (England)	821	6	94
Tropical Forest (Thailand)	211	58	42

[a] Data of Ovington (1962).

When a temperate forest is removed, the soil retains nutrients and structure and may be farmed for many years in the conventional manner, which involves plowing one or more times a year, planting short-season annual plants, and applying large amounts of quick-release inorganic fertilizers. During the winter, freezing temperatures help hold in nutrients and combat diseases and pests. In the humid tropics, on the other hand, forest removal takes away the land's ability to hold and recycle nutrients (as well as to combat pests) in the face of high year-round temperatures and periods of leaching rainfall. Such nutrients as are left are quickly drained away in the absence of organic or other holding mechanisms in the thin tropical soils. Crop production declines in a few years and the land is abandoned, creating the pattern of "shifting agriculture" (also called swidden agriculture) so common to the tropics. Thus, nutrient cycling in the temperate zone is more "physical" and in the tropics more "biological."

This brief account, of course, oversimplifies complex situations, but the contarst, as outlined above, underlies what is certainly the basic reason why sites in the tropics that support luxurious and highly productive forests yield so poorly under northern-style crop management. It is evident that a different type of agriculture needs to be designed for the humid tropics, perhaps one involving perennial plants with C_4 photosynthesis that also have symbiotic mycorrhiza (= root fungi) that facilitate direct recycling of nutrients (see next section). Unfortunately, agricultural science is so locked in with "northern" concepts, and so dependent on annual grains that the needed "revolution" is very slow in developing. For more on the theory of tropical cycling, see Went and Stark (1968).

RECYCLE PATHWAYS Since man will be concerned more and more with recycling problems it is instructive to summarize the subject of biogeochemistry in terms of recycle pathways. Figure 4-7 shows a number of routes by which nutrients and other substances are returned to the primary producers or other biotic levels. In any given large ecosystem all of these pathways are probably operating, but relative importance of each route is a function both of the type of ecosystem and the type of substance.

As already indicated recycling of many vital biogenic substances (nutrients) involves microorganisms and energy derived from the decomposition of organic matter. In ecosystems dominated by the detritus food chain, pathway 1 (Figure 4-7) is predominant. Where small plants, such as grass or phytoplankton, are heavily grazed, recycling by way of animal excretion may be important (path 2). In special cases an even more direct return (path 3) is accomplished by microorganisms that become a part of the plant, such as the mycorrhize fungi mentioned in the preceding section. This direct organism-to-organism cycle is not only important in the tropics, but also in the temperate zone where the soil (or water) is extremely poor in nutrients. Many conifers, such as pines and spruces, are able to grow well on extremely poor sites because of help they get from symbiotic microorganisms (for more on this see Wilde, 1968). As shown in the water cycle (Figure 4-6) many substances are recycled primarily by physical means involving solar or other nonbiotic energy (path 4, Figure 4-7). Finally, man enters the picture directly when he expends fuel energy to recycle water,nutrients (fertilizers) or metals (path 5). Note again that recycling requires energy dissipation from some source, such as organic matter (paths, 1, 2, 3), solar radiation (path 4), or fuel (path 5).

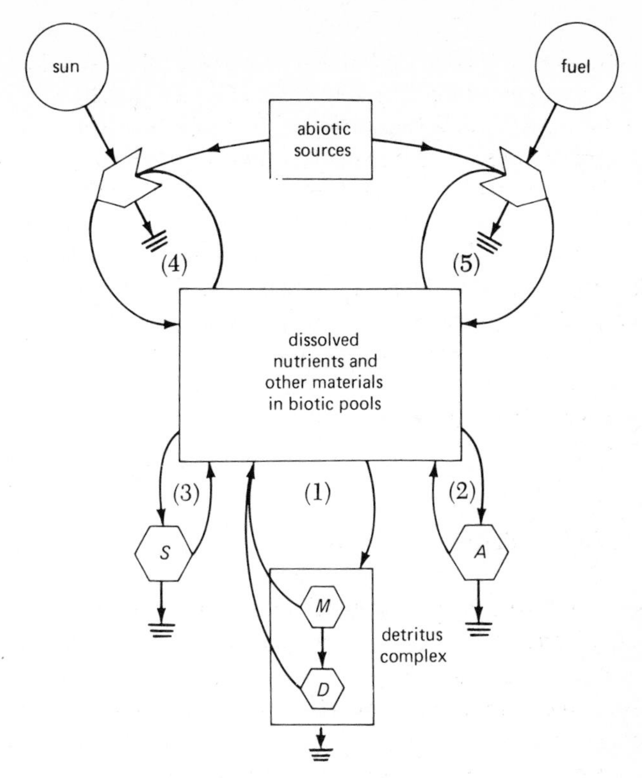

Fig. 4-7 Five major recycle pathways (1), (2), (3), (4), (5). A = animals: M = free-living microorganisms; S = symbiotic microorganisms; *D* = detritus consumers. Energy for pathways 1–3 comes from organic matter and for pathways 4 and 5 from solar or fuel energy. See text for further explanation.

LIEBIG'S LAW AND LIMITING FACTOR CONCEPTS

Any factor that tends to slow down the rate of metabolism or potential growth in an ecosystem is said to be a *limiting factor.* Where the brake, as it were, has survival value, as in the case of the herbivore limitation discussed in Chapter 3, the term *regulatory factor* may be more appropriate. The idea that organisms may be controlled by the weakest link in the ecological chain of requirements goes back a century or more to the time of Justus Liebig, who was a pioneer in the study of inorganic chemical fertilizers in agriculture.

Liebig was impressed with the fact that crop plants were often limited by whatever essential element was in short supply, regardless of whether the total amount required was large or small. Liebig's *law of the minimum* has come to mean that the rate of growth is dependent on the nutrient supplied or recycled in the minimum quantity in terms of need. If we extend this idea to include factors other than nutrients, and to include the limiting effect of the maximum (that is, too much can also limit), and recognize that factors interact (that is, short suppl of one thing affects requirements for another thing not in itself limit ing), we end up with a working principle that is very useful in th study of any specific ecosystem or any part thereof.

We may restate the extended *concept of limiting factors* as fol lows: The success of a population or community depends on a comple of conditions; any condition that approaches or exceeds the limit o tolerance for the organism or group in question may be said to be limiting factor. Although the quantity and quality of incoming energ and the laws of thermodynamics set the ultimate limits, differen ecosystems have different combinations of other factors that may pu further limitations on biological structure and function.

The chief value of the limiting factor concept lies in the fact tha it gives the ecologist an "entering wedge" into the study of complicate situations. Environmental relations are indeed complex, so it is fortu nate that not all factors are of equal ecological importance. Oxygen, fo example, is a physiological necessity to all animals, but it becomes limiting factor from the ecological standpoint only in certain environ ments. If fish are dying in a polluted stream, for example, oxygen con centration in the water would be one of the first things we woul investigate, since oxygen in water is variable, easily depleted, and ofte in short supply. If small mammals are dying in a field, however, w would look for some other cause, since oxygen in the air is constan and abundant in terms of need by the population (that is, not easil depleted by biological activity), and, therefore, not likely to be limitin to a population of air-breathing animals. We have already explaine why atmospheric CO_2 is a more important limiting factor for man tha oxygen, even though the latter is much more critical for man's interna physiology.

THE EXPERIMENTAL APPROACH TO THE STUDY OF LIMITING FACTORS

The prefix "eury" is often used to indicat wide limits of tolerance and "steno" to ind cate narrow limits. Among fish, trout are i general more stenothermal than bass in tha they are not able to tolerate as wide a rang of temperatures. If we cut down all the trees along a mountain strean allowing the sun to warm up the water a few degrees, the trout migh be killed and the bass persist. Organisms with wide limits of toleranc of course, are likely to be widely distributed, but wide limits for on factor does not necessarily mean wide limits for all factors. A pla might be eurythermal but stenohydric (have narrow limits of toleranc for water); or an animal such as a trout might be stenothermal but euryphagic (feed on a wide variety of food).

Limits of physiological tolerance for such things as temperature or nutrients can often be determined with precision in the laboratory, but we should be cautious about transferring such knowledge to the field. As we have previously pointed out in our discussion of the integrative levels concept (Chapter 1), what is true at the organism level may be only part of the story at the community level. Quite frequently organisms do not actually live in nature under optimum conditions for a specific factor since other factors or interactions may be more important for survival within the ecosystem.

Studies at sea provide examples of experimental approaches that combine field and laboratory techniques. In the photic (lighted) zone of the open sea, most nutrients are relatively scarce and therefore are limiting factor suspects. However, when so many nutrients are in low concentration, the chemical observations normally made at sea do not provide any information as to which nutrient is actually limiting the production at a particular time or place. Simple enrichment experiments may help answer the question. The experiments entail adding various nutrients one by one to water containing the natural phytoplankton, placed in containers on board ship or suspended in bottles in the sea. Any nutrients that increase the rate of photosynthesis is judged to be limiting at the time the sample was taken. When Ryther and colleagues (Ryther and Guillard, 1959; Ryther and Dunstan, 1971) added nitrogen or phosphorus to the water from the Sargasso Sea, which is one of the more or less desert areas of the ocean mentioned in Chapter 3, no increase in photosynthesis occurred (as compared with control samples not enriched), even though these nutrients are so commonly limiting in the sea. When either silica or iron was added, however, photosynthesis immediately increased, thus indicating that the micronutrients were either directly limiting or in some way necessary for full use of the macronutrients. In other words, adding a lot of nitrate and phosphate fertilizer in the area would not increase food production unless the silicate and iron limitations were also overcome by enrichment with these nutrients.

LIMITING FACTORS IN STEADY-STATE D TRANSIENT SYSTEMS

Liebig's law is most applicable in steady-state conditions and least applicable under transient conditions (see Figure 1-3 for an explanation of these ecosystem "states"). When inflows balance outflows in the steady-state, one or a few factors may well control the rate of function. For example, the rate of release of carbon dioxide from decaying organic matter often controls the rate of production in a body of water that is in steady-state. Should the ecosystem be perturbed out of equilibrium as, for example, by a storm that brings in more CO_2 and changes other conditions, then during the period of readjustment, or transition, there may be no minimum factor; instead the rate of function will depend on rapidly changing concentrations and interactions of many factors.

Water pollution control specialists have, in the past, failed to understand these relationships and have attempted to single out one factor as the cause of undesirable, but unstable, algal blooms resulting from pollution (the one-problem/one-solution syndrome previously mentioned). For a while phosphorous was a favorite culprit, but a ban on phosphates in detergents did not solve most problems because nitrogen, carbon dioxide, and many other constituents, along with phosphorus rapidly replace one another as growth-promoting factors in the transitory oscillations that characterize most noxious "blooms" of algae. There is no theoretical basis for any "one-factor" control hypothesis under transient-state conditions. The strategy of pollution control must involve reducing the input of all enriching and toxic materials, not just one or two items.

FACTOR COMPENSATION

Caution is necessary not only in transferring laboratory data on tolerances to the field but in transferring such data from one region to another, even when the same species are involved. Species with wide geographical ranges often develop locally adapted populations, called *ecotypes*, having different limits of tolerance to temperature, light, or other factors. Compensation along a gradient of conditions may involve genetic races (with or without morphological manifestations) or merely acclimatization. Reciprocal transplants often reveal whether ecotypes are actually genetic races. The possibility of genetic fixation in local strains has often been overlooked in applied ecology, with the result that restocking or transplanting has often failed because individuals from remote regions were used.

A good example of temperature compensation within the species is shown in Figure 4-8. Small jellyfish move through the water by rhythmical contractions that expel water from the central cavity in a sort of jet propulsion. A pulsation rate of about 15 to 20 per minute seems to be optimum. Note that individuals living in the northern sea at Halifax swim at approximately the same rate as individuals in southern seas, even though the water temperature may be 20°C lower.

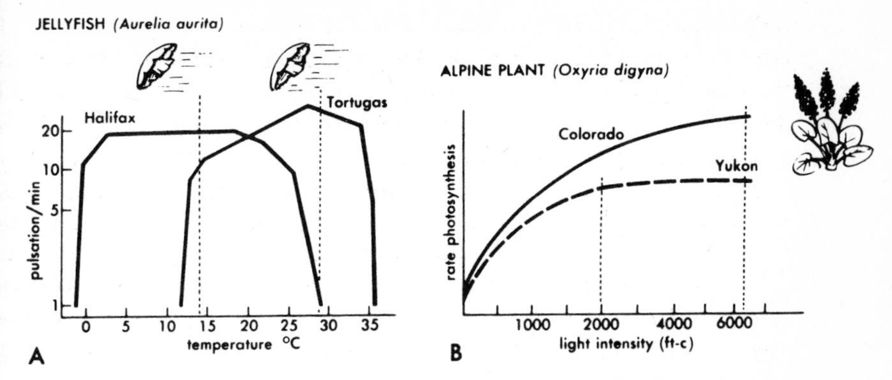

Fig. 4-8 Physical factor compensation by animals and plants. (A) How different populations of the same species (i.e., ecotypes) of jellyfish are adapted to swim at temperatures of their environment. The dotted vertical lines represent the average water temperature in summer at the northern (Halifax) and southern (Tortugas) locations. (B) How populations of an alpine plant in the Yukon, where light intensity is low, reach maximum photosynthetic rate at a lower light intensity (about 2000 foot candles, as indicated by the first dotted line) than plants in high mountains of Colorado where light intensity is high. (*Left,* redrawn from T. H. Bullock, *Biological Reviews,* Vol. 30, 1955, after Mayer. *Right,* redrawn from H. A. Mooney and W. D. Billings, *Ecological Monographs,* Vol. 31, 1961.)

Figure 4-8 illustrates partial light compensation by plant populations. Alpine grasslands at low latitudes and in the arctic may be subject to similar low temperatures but to very different light intensities. As shown in the diagram, the plants in the Yukon Territory reach their peak of photosynthesis (that is, light saturation) at a lower light intensity than the more southern alpine populations in Colorado, in harmony with the generally lower intensity in the former region. Recent discoveries in photosynthetic mechanisms that adapt plants to differing light intensities have already been discussed (page 77).

Although many species are able to compensate along an extensive gradient, such as a north-south temperature gradient, more complete adaptation is often accomplished from the ecological standpoint by a series of closely related species that replace one another along the gradient. For example, along the seashore in New England small white snails, or periwinkles, are abundant in the intertidal zone; *Littorina littorea* is usually the most common species. Further south, along the South Atlantic coast, this species is replaced in the same intertidal zone by a similar but distinct species, *Littorina irrorata.* Still other closely related species, *L. planaxis, scutulata,* and *ziczac* are found on the west coast and on down into the tropics. So just about on any rocky coast you will find some kind of periwinkle. We have already remarked on the importance of such ecological equivalence in Chapter 2. In many ways interspecific compensation of this sort is more efficient than intraspecific compensation since a wider range of genetic material is available.

ECOLOGICAL INDICATORS Despite the wide range of adaptation, certain species may sometimes serve as useful indicators of environmental conditions. From what we have just said it is obvious that "steno" species make more reliable indicators than "eury" species. Therefore, the rarer species often make the best indicators. Range managers, for example, find that the decline of certain relatively rare species of plants that are sensitive to grazing will indicate the approach of overgrazing before it becomes apparent in the grassland as a whole. A group of species or the whole community, of course, provides the best indicator of conditions, although harder to assess. Investigators studying stream pollution have found that a decline in the number of species or in a diversity index (as discussed in Chapter 2) often indicates pollution before the total number of individuals or the total productivity are measurably affected. Employing the species structure or the diversity ratio as a sensitive index may enable the applied worker to recognize and correct a situation involving a limiting factor before it becomes critical.

To summarize, there are three important points regarding limiting factors to be kept constantly in mind: (1) Coordinated field observation, field experimentation, and laboratory experimentation are almost always necessary in the investigation of limiting factors at the ecological level. (2) The presence of a species in two different regions does not necessarily mean that the environmental conditions are the same in the two regions; although rare species or "steno" species sometimes make good indicators, groups of species or the community as a whole is more reliable. (3) Communities are able to adapt or compensate so that the overall rates of function, such as energy flow and productivity, may remain the same over a considerable part of a gradient of conditions even though the species structure may change drastically. Thus, within the middle range of temperatures in the biosphere, northern communities may be able to fix and transfer as much energy as southern communities on a yearly average (assuming, of course, that conditions other than temperature are not too extreme). Within fairly wide limits therefore, differences in conditions of existence have more effect at the species level than at the ecosystem level. When the conditions do approach the extremes of a gradient, the number of species declines first, followed by decline in the *rates of function* as the conditions become limiting for any and all life (as in deserts or in the arctic). As pointed out in Chapter 2, stability also is affected in that the standing crop and the rate of function fluctuate or oscillate more violently than is the case under the moderate range of conditions of existence.

[CONDI]TIONS OF EXISTENCE [AS R]EGULATORY FACTORS Physical conditions of existence may not only be limiting factors in the detrimental sense but also regulatory factors in the beneficial sense that adapted organisms respond to these factors in such a way that the community achieves the maximum homeostasis possible under the conditions. Discussion of the many mechanisms by which organisms adapt is beyond the scope of this volume; an example or two will serve to illustrate the extensive role of physical factors in coordinating biological activity.

No physical factor is of greater interest to the ecologist than light. It is, first, a source of energy; second, a limiting factor (since too little or too much kills); and, third, an extremely important regulator of daily and seasonal activities for a great many organisms, both plant and animal. Three aspects of light are of great interest: (1) the intensity, as illustrated in Figure 4-8B; (2) the wavelength; and (3) the duration. As shown in Figure 4-9, visible light is but a small part of an

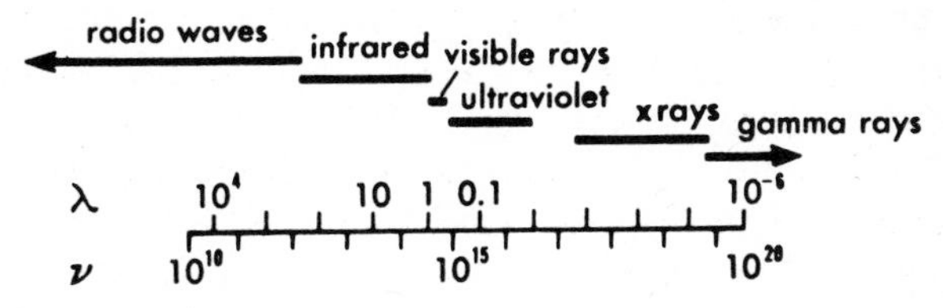

Fig. 4-9 The spectrum of electromagnetic radiations. λ = wavelength in microns; v = frequency in seconds.

extensive radiation spectrum of electromagnetic energy ranging from very short to very long wavelengths. Infrared, ultraviolet, and x rays also have ecological importance, but our attention here will be on the visible rays. One of the most dependable environmental cues by which organisms time their activities in temperate zones is the day-length period, or photoperiod (Figure 4-10). In contrast to temperature. photoperiod is always the same for a given season and locality, year after year. Photoperiod has been shown to be the timer or trigger that sets off a physiological sequence that brings about molting, fat deposition, migration, and breeding in temperate-zone birds. Not all of the

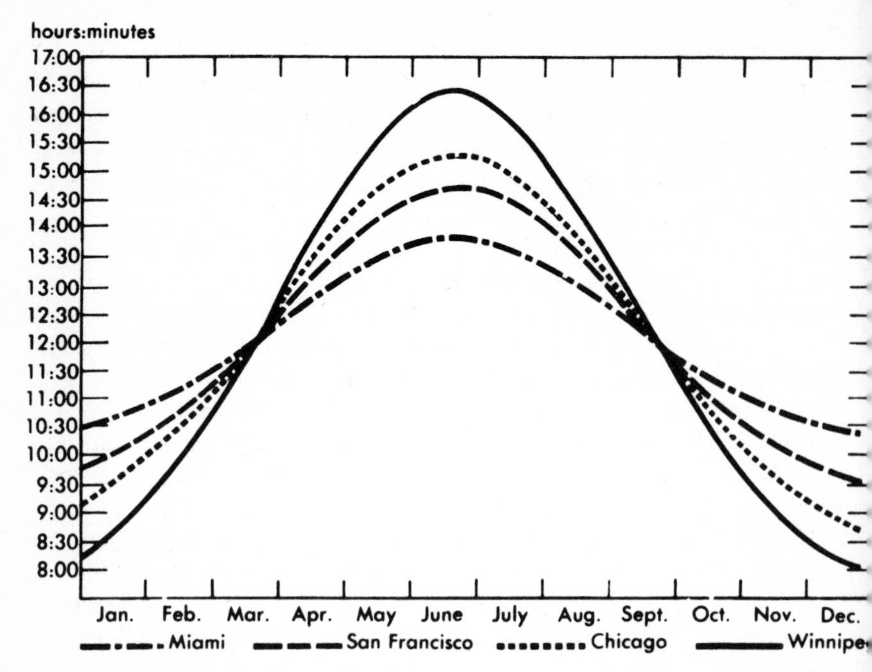

Fig. 4-10 Variations in length of day or photoperiod at four latitudes i North America. Photoperiod is an important regulator of seasonal activit in both plants and animals, especially in north temperate regions where sea sonal changes in day length are pronounced. The cocklebur requires a shor day (15 hours or less) and hence does not flower at Winnipeg until the day begin to shorten in late summer. At Miami, however, the days are alway short enough so that the plant flowers whenever physiologically ripe. (Re drawn from F. B. Salisbury. Copyright © by *Scientific American,* Ma 1952. All rights reserved.)

details have been worked out, but our present understanding of photo period regulation of the annual cycle in birds is now fairly well under stood. On the wintering ground in autumn the bird is refractory t photoperiod stimulus; that is, the internal mechanism will not respon to a day-length stimulus. Short days of fall are apparently necessar to reset and rewind the biological clock, as it were. Any time afte December the bird will respond to an increase in day-length. In nature of course, stimulation does not occur until the approach of the lon days of late winter or spring. However, one can produce out-of-seaso fat deposition, migratory restlessness, and an increase in size of repro ductive organs in midwinter in the laboratory by an artificial increas in the light period. Interestingly enough, the intensity of the extra ligh is not a factor so long as the light is stronger than full moonlight; thus the adaptation neatly avoids false stimulation by moonlight on the on hand, and the effects of varying daylight intensity due to cloud cove on the other.

Photoperiod is believed to act through the hypothalamic part of the brain (by way of the eye), which produces a neurohormone that stimulates the pituitary gland (the master endocrine gland), which in turn sends out the several different hormones in sequence to the target organs. The whole sequence can be likened to a seasonal biological clock that is set and regulated by length of day.

Among other seasonal activities that have been shown to be under photoperiod control are: flowering in many plants, elongation of stems in seedling plants, hibernation and seasonal change in hair coats in mammals, reproduction in many animals other than birds, and diapause (resting stage) in insects. In plants, day-length acts through a pigment called *phytochrome*, which in turn activates enzymes controlling growth and flowering processes. It has even been shown that the number of underground nitrogen-fixing root nodules in legumes is controlled by photoperiod acting through the leaves of the plant. Since nitrogen-fixing bacteria in the nodules require food energy from the plant if they are to do their work, the more light and chlorophyll the more food for the bacteria; maximum coordination between the plant and its bacterial partners seems to be enhanced by the photoperiod regulator.

Insects that have two generations a year have evolved a neat day-length response that, as in birds, involves a neurohormone mechanism. Long days stimulate the "brain" (actually a nerve cord ganglion) to produce a hormone bringing on a diapause or resting egg, but short days do not have such an effect. Thus, individuals of the first generation hatched during the short days of spring produce eggs that immediately hatch into a second generation. Individuals in the second generation subjected to the long days of summer produce diapause eggs that do not hatch until the next spring, no matter how favorable temperature, moisture, and other conditions may be.

Light, of course, is not the only regulator. Seeds of many annual desert plants sprout only when there is a shower of a certain minimum magnitude (½ in. of rain or more, for example); the mechanism here seems to involve a chemical germination inhibitor that must be washed out of the seed coat. These few examples will serve to emphasize the reciprocal relationships between organisms and their nonliving environment.

THE SPECIAL CASE OF FIRE AS AN ECOLOGICAL FACTOR

Contrary to popular opinion, fire in nature is not a completely artificial factor created by man, nor is it always detrimental to man's interests. Fire is an important environmental factor in many terrestrial ecosystems and was important long before man attempted to control it. Because man can, within limits, control fire, it is especially important that he study this factor thoroughly, and with an objective mind. If we cannot learn to handle this relatively simple environmental factor in our own best interest, we have no business attempting to control rainfall or other vastly more complex matters.

Fire is both a limiting and a regulatory factor, as are most of the other factors we have discussed. It is important in warm or dry regions, and regions with warm and dry seasons such as in the southern third of the United States or in Central Africa. In such areas seasonal or periodic light fires apply selective pressure that favors the survival and growth of some species at the expense of others. Many natural communities in such regions are "fire types" in that their prosperity or very survival depends on fire. The effect of fire in one such community is shown in the sketches in Figure 4-11. In this case the grass is not only adapted to fire but is much more valuable to man than the desert shrubs that tend to increase in the absence of fire. If man wants to keep fire out of such communities he must substitute something else in order to prevent a change to an economically less desirable vegetation. Chemicals may be effective, but they are much more expensive than "controlled" or "prescribed" burning as practiced, for example, in southeastern long-leafed pine forests, another fire-type ecosystem.

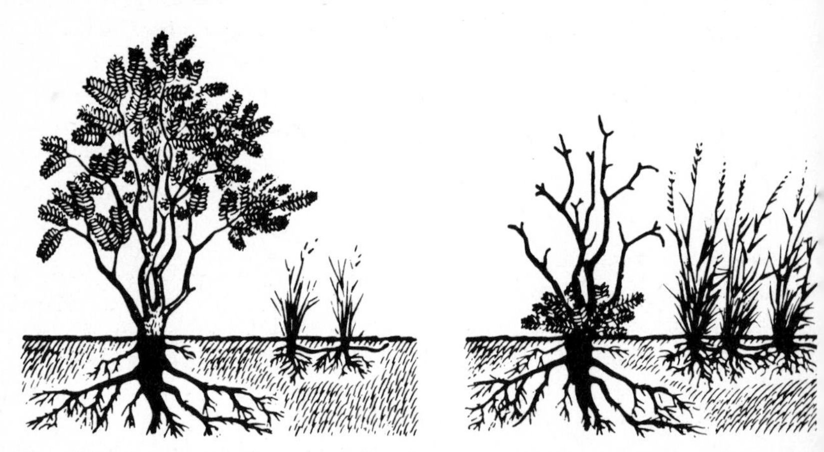

Fig. 4-11 Diagrams show how fire favors grass over mesquite shrubs in southwestern U.S. In the absence of fire the mesquite chokes out the grass (left). After a fire, grass recovers quickly, growing with increased vigor under conditions of reduced competition (right). Controlled burning will eliminate the mesquite entirely and maintain the grassland. (Redrawn with permission. Copyright © 1961 by *Scientific American,* Inc. All rights reserved.)

Although more work needs to be done in this area, it now appears that in dry or hot regions fire acts as a decomposer to bring about a release of mineral nutrients from accumulated old litter that becomes so dry that bacteria and fungi cannot act on it. Thus, fire may actually increase productivity by speeding up recycling. Certainly the big game herds of Africa or the deer in California chaparral (a fire-type shrub) do not thrive unless periodic fires bring on a flush of new palatable grass or foliage. Furthermore, periodic light fires prevent the start of bad fires by keeping the combustible surface litter to a minimum. In southern California, for example, fire prevention in the chaparral vegetation has often resulted in severe fires that wipe out many homes. An example of the interaction of fire and grazing is briefly described on page 193.

It is extremely important that we distinguish between the light surface fires characteristic of the fire-type ecosystem and the wild forest fires of northern forests; the latter, of course, are all bad, since they destroy nearly the entire community. Because man, by his carelessness, tends to increase such holocausts, it is necessary that the public be made sharply aware of the necessity of fire prevention in forests. The intelligent citizen should recognize that he, as an individual, should never start or cause fires anywhere in nature; however, the scientific use of fire as a tool by trained persons is quite a different matter in the natural environment. Fire is an inexpensive form of energy dissipation capable of doing much good or much harm, depending on intensity, timing, and the adaptive nature of the natural community.

SUGGESTED READINGS

References cited

Berkner, L. V. and L. C. Marshall. 1965. History of major atmospheric components. *Proc. Natl. Acad. Sci., Washington, D.C.* 53:1215–1226. Also in *The Origin and Evolution of Atmospheres and Oceans,* ed. Brancazio and Cameron, pp. 102–120. New York: John Wiley & Sons.

———. 1966. The role of oxygen. *Saturday Rev.* 10th Anniversary Issue, May 7, 1966, pp. 30–33.

Bormann, F. H. and G. E. Likens. 1967. Nutrient cycling. *Science.* 155: 424–429.

———. 1970. The nutrient cycles of an ecosystem. *Sci. Amer.* 223(4): 92–101.

Hanson, W. C. and Kornberg, H. 1956. Radioactivity in terrestrial animals near an atomic energy site. *Proc. Int. Conf. Peaceful Uses Atomic Energy, Geneva.* 13:385–388.

Hutchinson, G. E. 1948. On living in the biosphere. *Sci. Monthly.* 67:393–398.

Likens, G. E.; F. H. Bormann; and N. M. Johnson. 1969. Nitrification: importance to nutrient losses from a cut-over forested ecosystem. *Science.* 163:1205–1206. See also *Ecol. Monogr.* 40:23–47, 1970 and *Science.* 159:882–884, 1968.

Likens, G. E. and F. H. Bormann. 1972. Nutrient cycling in ecosystems. In *Ecosystems: Structure and Function*, ed. J. A. Wiens, pp. 25–67. Oregon State Univ. Press.

________. 1974. Acid rain: a serious regional environmental problem. *Science.* 184:1176–1179. See also: *Environment.* 14(2):33, 1972.

Odum, H. T. 1970. Energy values of water resources. *Proc. 19th Southern Water Res. and Poll. Control Conf. Duke Univ. Press*, pp. 56–64.

Ophel, Ivan L. 1963. The fate of radiostrontium in a freshwater community. In *Radioecology*, eds. V. Schultz and A. W. Klement, pp. 213–216. New York: Reinhold.

Ovington, J. D. 1962. Quantitative ecology and the woodland ecosystem concept. *Adv. Ecol. Res.* J. B. Cragg, ed. 1:103–192. New York: Academic Press.

Plass, Gilbert N. 1959. Carbon dioxide and climate. *Sci. Amer.* 201(1):41–47.

Pomeroy, L. R. 1970. The strategy of mineral cycling. *Ann. Rev. Ecol. and Systematics.* 1:171–190.

Redfield, Alfred C. 1958. The biological control of chemical factors in the environment. *Amer. Sci.* 46:205–221.

Ryther, J. H. and R. R. Guillard. 1959. Enrichment experiments as a means of studying nutrients limiting to phytoplankton production. *Deep-Sea Res.* 6:65–69.

Ryther, J. H. and W. M. Dunstan. 1971. Nitrogen, phosphorus and eutrophication in the coastal marine environment. *Science.* 171:1008.

Went, F. W. and N. Stark. 1968. Mycorrhiza. *Bio-Sci.* 18:1035–1039.

Wilde, S. A. 1968. Mycorrhizae and tree nutrition. *Bio-Sci.* 18:482–484.

Woodwell, G. M. 1967. Toxic substances and ecological cycles. *Sci. Amer.* 216(3):24–31.

Woodwell, G. M.; C. F. Wurster; and P. A. Isaacson. 1967. DDT residues in an east coast estuary: A case of biological concentration of a persistent insecticide. *Science.* 156:821–824.

Biogeochemical cycles

Alexander, Martin, ed. 1972. Accumulation of nitrate. *Nat. Acad. Sci Washington, D.C.* (Good review of the nitrogen cycle and a discussion of possible harmful effect of nitrate accumulation in soils, water, and food as result of man's heavy use of nitrogen fertilizers.)

Bolin, Bert. 1970. The carbon cycle. *Sci. Amer.* 223(3):124–132.

Cloud, Preston and Aharon Gibor. 1970. The oxygen cycle. *Sci. Amer.* 223(3):110–123.

Cole, L. C. 1966. Protect the friendly microbes. *Saturday Review*. 10th Anniversary Issue, May 7, 1966, Vol. 49, p. 46. (Essay on role of microorganisms in regeneration of nutrients.)

Deevey, Edward S., Jr. 1970. Mineral cycles. *Sci. Amer.* 223(3):148–158.

Delwiche, C. C. 1970. The nitrogen cycle. *Sci. Amer.* 223(3):136–146.

Frieden, Earl. 1972. The chemical elements of life. *Sci. Amer.* 227(1):52–60.

Janzen, Daniel H. 1973. Tropical agroecosystems. *Science*. 182:1212–1219.

Kellogg, W. W.; R. D. Cadle; E. R. Allen; A. L. Lazrus; and E. A. Martell. 1972. The sulfur cycle. *Science*. 173:587–596. (Man is now contributing about one half as much as nature to the total atmospheric burden of sulfur compounds. In industrial regions he is overwhelming natural removal processes.)

Penman, H. L. 1970. The water cycle. *Sci. Amer.* 223(3):98–108.

Air pollution and its effect on climate and human health

Bryson, Reid A. 1974. A perspective on climatic change. *Science*. 184:753–760. (Marshals evidence to show that small changes in such atmospheric variables as turbidity and CO_2 content can alter climates and that these variables are indeed being altered by man.)

Cutchis, P. 1974. Stratospheric ozone depletion and solar ultraviolet radiation on earth. *Science*. 184:13–19. (Calculates the extent that a hypothetical fleet of stratospheric aircraft would reduce the ozone shield that protects life from excessive ultraviolet radiation.)

Lal, D. and Hans E. Suess. 1968. The radioactivity of the atmosphere and hydrosphere. *Ann. Rev. Nuclear Sci.* 18:407–434.

Landsberg, H. E. 1970. Man-made climatic changes. *Science*. 170:1265–1274.

Lave, Lester B. and E. P. Seskin. 1970. Air pollution and human health. *Science*. 169:723–733.

Lowry, W. P. 1967. The climate of cities. *Sci. Amer.* 217(2):15–23. (Large cities have less sunshine, more fog and drizzle, and higher air temperatures than surrounding countryside.)

Newell, R. E. The global circulation of atmospheric pollutants. 1971. *Sci. Amer.* 224(1):32.

Limiting factor concepts

Browne, C. A. 1942. Liebig and the law of the minimum. In Liebig and after Liebig, ed. F. R. Moulton. *Publ. Amer. Assoc. Adv. Sci.* 16:71–82.

Bullock, T. H. 1955. Compensation for temperature in metabolism of poikilotherms. *Biol. Rev.* 30:314–342.

Hutchinson, G. Evelyn. 1973. Eutrophication. *Amer. Sci.* 61:269–279. (Reviews limiting factor concepts as applied to the enrichment of lakes by man.)

Odum, E. P. 1971. *Fundamentals of Ecology,* 3rd ed. Chapter 5. Philadelphia: Saunders.

Photoperiodism

Beck, S. D. 1960. Insects and the length of day. *Sci. Amer.* 202(2):108–118.

Butler, W. L. and R. J. Downs. 1960. Light and plant development. *Sci. Amer.* 203(6):56–63.

Evans, L. T., Ed. 1963. *Environmental Control of Plant Growth.* New York: Academic Press.

Farner, D. S. 1964. The photoperiodic control of reproduction in birds. *Amer. Sci.* 52:137–156.

Soils

Russell, E. W. 1961. *Soil Conditions and Plant Growth,* 9th ed. New York: Longman.

Richard, B. N. 1974. *Introduction to the soil Ecosystem.* New York: Longman. (Excellent accounts of functional processes including chapters on decompositions and eriergy flow; microbiological processes and nutrient cycling; the rhizospheres; mycorrhiza; root nodule symbiosis and nitrogen cycle.)

Fire as an ecological factor

Ahlgren, I. F. and C. E. Ahlgren. 1960. Ecological effects of forest fires. *Bot. Rev.* 26:483–533.

Cooper, Charles F. 1961. The ecology of fire. *Sci. Amer.* 204(4):150–160.

Dodge, Marvin. 1972. Forest fuel accumulation—a growing problem *Science.* 177:139–142. (A forest ranger concludes that complete forest fire protection is impossible and undesirable ecologically. Prescribed low-intensity burning often necessary to reduce dead material that fuels high-intensity fires.)

Muller, C. H.; R. B. Hannawalt; and J. K. McPherson. 1968. Allelopathic control of herb growth in the fire cycle of California chaparral. *Bull Torrey Bot. Club.* 95:225–231. (Interaction of antibiotics and fire in a "fire-type ecosystem.")

Oberle, Mark. 1969. Forest fires: suppression policy has its ecological drawbacks. *Science.* 165:568:571. (A staff writer's review of expert opinions on undesirable side effects of fire suppression and on use of "prescribed burning," that is, the use of fire to fight fire.)

chapter 5

Population Ecology

We now come to the more purely biological aspects of ecology, that is, the interaction of organisms with organisms in the maintenance of community structure and function. Up to this point much of our attention has been focused on the role of the great physical and chemical forces in the ecosystem. We have outlined how energy from the sun flows through natural and seminatural ecosystems, such as the oceans, croplands, and forests, and we have seen how fuel-energy flows guided by man supplement, modify, and interact with the solar-powered biosphere. Likewise, we have demonstrated how materials are cycled and recycled, and how populations and communities are adapted and limited by temperature, light, nutrients, and other abiotic factors. We have already emphasized that organisms are not just pawns in a great chess game in which the physical environment directs all the moves. Quite to the contrary, natural communities, as well as man, modify, change, and regulate their physical environment within certain limits; we have cited a number of examples in the previous chapters, as, for instance, bacterial regeneration of nitrogen, the oxygenation of the atmosphere, or the reef-building activities of cooperating teams of corals and algae. In Chapter 2 we considered in some detail diversity as an index to numerical relationships of populations in the community matrix. What we have not yet considered in detail are principles dealing with interactions within and between populations. Population ecology is of special interest to biologists because the population is the principal, but not the only unit involved in evolution. Thus, population ecologists and population geneticists share common interests.

POPULATION ATTRIBUTES A population, as you recall from Chapter 1, is defined as the collective group of a particular kind in the community. In practice, a population is simply all of the organisms of the same species found occupying a given space. A population, as with any other level of organization, has a number of important group properties not shared by adjacent levels (the organism, on the one hand, and the community on the other; see Figure 1-1). The more important of these population characteristics, or group attributes, are as follows:

Density: population size in relation to a unit of space.

Birth rate, or more broadly, natality (so as to include organisms that arise from seeds, spores, eggs, and so on): the rate at which new individuals are added to the population by reproduction.

Death rate or *mortality*: the rate at which individuals are lost by death.

Dispersal: the rate at which individuals immigrate into the population and emigrate out of the population.

Population growth rate or *growth form*: the net result of natality, mortality, and dispersal.

Dispersion: the way in which individuals are distributed in space, generally in one or more of the following three broad patterns: (1) random distribution, in which the probability of an individual occurring in any one spot is the same as the probability of it occurring at any other spot; (2) uniform distribution, in which components occur more regularly than random, such as corn in a cornfield; (3) clumped distribution (the most common in nature), in which individuals or other components are more irregular than random, as for example, a clump of plants arising from vegetative reproduction, a flock of birds, or people in a city.

Age distribution: the proportion of individuals of different ages in the group.

Genetic characteristics: especially applicable to population ecology, as for example, adaptiveness, reproductive (Darwinian) fitness, and persistence (that is, probability of leaving descendents over long periods of time).

Individual organisms are born, die, have age, and grow, but such characteristics as birth rate, death rate, density, and the others listed above are meaningful only at the group level. If we are to understand thoroughly the ecology of a species, we must study and measure these population group characteristics as well as know the life history and identifying features of the species. And, of course, we all must be concerned about the vital statistics of the human population.

It is quite evident that the degree of crowding, or density, and the pattern of dispersion of individuals (whether random, uniform, or clumped together in a limited area) are especially important in determining the degree of interaction between individuals of the same or other species. Not so evident is the role that the type of population growth plays in the response of species to each other and to the factors of the physical environment. In this brief introduction to population dynamics we will concentrate on growth form, leaving detailed consideration of the other attributes listed above for more advanced texts.

POPULATION GROWTH FORM

Figure 5-1 shows two contrasting ways that populations grow when opportunity presents itself; for example, at the beginning of a growing season, when a few individuals enter (or are introduced into) an unoccupied area, or when unused resources become available. For convenience we can label these as the *J-shaped growth form* (Figure 5-1A) and the S-*shaped* or *sigmoid growth form* (Figure 5-1B). In the former, the population density increases in exponential or geometric fashion; for example, 2, 4, 8, 16, 32 . . . , and so on, until the population runs out of some resource or encounters some other limitation (limit N, Figure 5-1A). Growth then comes to a more or less abrupt halt and den-

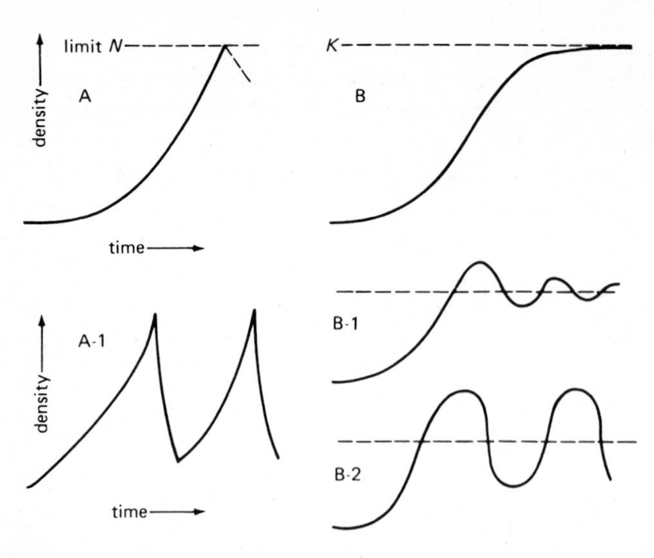

Fig. 5-1 Two types of population growth form the J-shaped (exponential) (A), and the S-shaped (sigmoid) (B), and some variants. A-1 illustrates severe oscillations that would result from "boom and bust" cycles of exponential growth. B-1 and B-2 show dampened and undampened oscillations that occur when sigmoid growth overshoots the carrying capacity, *K*.

sity usually declines rapidly until conditions for another rapid growth episode are restored (A-1 in Figure 5-1). Populations with this kind of growth form are unstable unless regulated by factors outside of the population. In the S-shaped growth pattern (Figure 5-1B) self-limiting or self-crowding negative feedback reduces the rate of growth more and more as density increases. If the limitation is linearly proportional to density we get a symmetrical sigmoid curve with density levelling off so as to approach an upper asymptote level, *K*, commonly called the carrying capacity level because it represents the maximum sustainable density. This pattern enhances stability since the population regulates itself. However, in the real world, density often overshoots carrying capacity because of time lags in feedback control resulting in oscillations, two types of which are shown as B-1 and B-2 in Figure 5-1. If oscillations decrease in amplitude with time, as in B-1, then for all practical purposes a steady-state will be achieved. All these contrasting growth forms may be combined or modified, or both, in various ways according to the growth potential of the species and the properties of the ecosystem.

In Figure 5-2 the temporarily unlimited and the self-limited growth forms are plotted with density on a logarithmic rather than arithmetic scale. This type of graph is called a semilog plot since one axis (density in this case) is log and the other remains arithmetic (time in this case). This type of plot has the advantage that the

J-shaped growth curve becomes a straight line whose slope represents the growth rate constant. The sigmoid growth form becomes a convex curve showing how growth rate decreases with density until it is zero. The slope of the tangent at any point represents the rate of growth at that point in time. As long as growth continues as a straight line on a semilog plot it can be said to be exponential. It is evident that exponential growth, whether in a population or involving something like man's consumption of fuel, cannot be allowed to continue for long without great danger of a disastrous overshoot because with each doubling time the jump becomes larger and larger. Thus, if a leaf-eating population of insects in a tree should increase unrestricted at a rate of tenfold each month, there might be only 100 individuals after two months, but 10,000 after four months, enough to quickly strip the tree bare of leaves.

For readers with a mathematical background, formulas for the two types of curves are given in Figure 5-2. For now we need only call attention to two important constants in the equations, namely, *K*, as already noted, and *r*, which represents the basic or intrinsic rate of growth of the population in an unlimited environment. We shall have occasion to refer to these two properties in the next chapter.

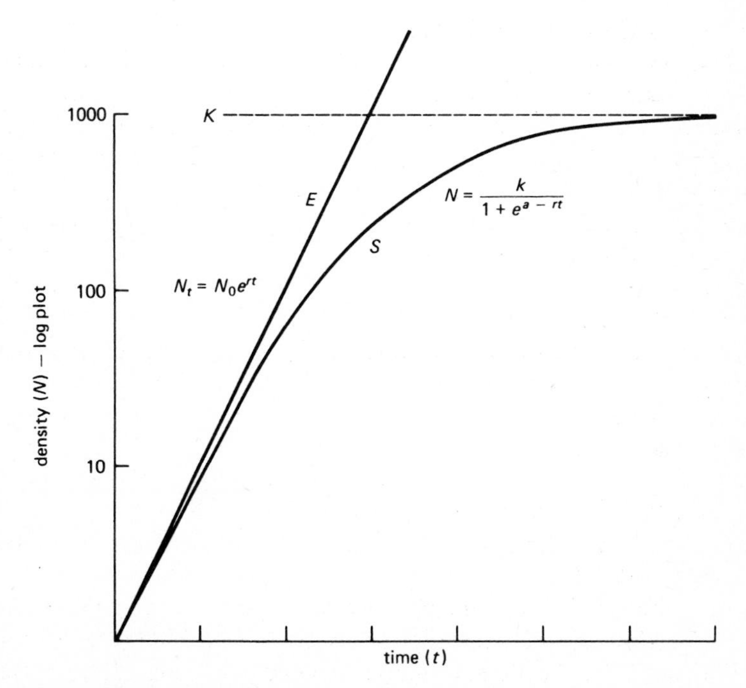

Fig. 5-2 Exponential and sigmoid growth plotted on a semilog graph (density on a logarithmic, time on an arithmetic scale). See text for explanation.

In summary, some populations tend to be self-limited in that the rate of growth decreases as the density increases. Such populations tend to level off in density before saturation, and their population growth can be said to be inversely *density dependent*. Other populations are not self-limited but tend to grow exponentially until checked by forces outside of the population (that is, other populations or general ecosystem limitations); such populations may overshoot their energy and habitat resources, literally crowding themselves to death, in the case of plants; or eating themselves out of food and home, in the case of animals. Their population growth can be said to be *density independent*, at least until the density becomes very great. When poorly regulated by factors outside of the population itself, species of the latter type are subject to severe oscillations in density and may become serious pests to man. In fact, a pest can be defined as an opportunist capable of rapid exponential growth when control within the ecosystem breaks down. We cited examples of this in Chapter 1.

There is still a third type of relationship between density and growth rate. In some species the reproductive rate is greater at intermediate density than at either low or high density; in other words, both "undercrowding" and "overcrowding" are limiting. Such a pattern is called the *Allee growth type* after the late W. C. Allee (1961). Some species of sea gulls are good examples. Most sea gulls nest in colonies on isolated islands or other protected places. In certain species it has been shown that the number of young produced per pair is greater when the density of the breeding colony is fairly large than when only a few pairs are present or when there is severe crowding. In highly social species, behavior patterns necessary for pair formation and efficient care of the young are apparently stimulated and augmented by the nearby presence of other individuals. Oysters are also more successful when the local density is moderately high, but for another reason. The oyster begins its life as a planktonic larvae that must settle on some suitable hard substrate if it is to survive. Where a large number of old oysters are present in a colony their shells provide a favorable place for the settling of the larvae. When man greatly reduces the oyster colonies by "overfishing," population growth becomes so low that the colony recovers very slowly (or perhaps not at all), even though no longer "fished" and even though there are planktonic larvae present in the water. Thus, conservation of oysters or other species that exhibit low growth rate at low density depends on maintaining a large standing crop at all times.

The three patterns of growth rate in relation to density are diagrammed in Figure 5-3. Note that the growth rate per unit of population (that is, per individual, pair, 100 individuals, or other convenient

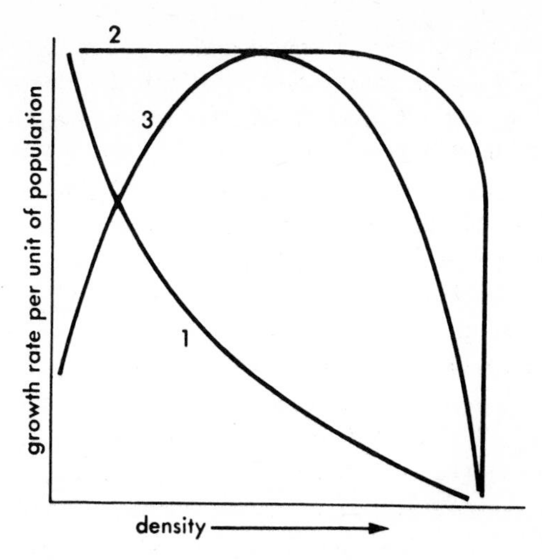

Fig. 5-3 Three patterns of population growth rate in relation to population density. (1) The growth rate decreases as density increases (self-limiting or inverse density-dependent type). (2) Growth rate remains high until density become high and factors outside of the population become limiting (density-independent type). (3) Growth rate is highest at intermediate densities (the Allee type).

unit), and not the total growth rate, is plotted against total density. Thus, in the inverse density-dependent or self-limiting type the number of individuals added to the population per individual per unit of time decreases as the density increases (curve 1 in Figure 5-3). In the density-independent or nonself-limiting type the growth rate per individual continues at a high level until a high density is reached (curve 2 in Figure 5-3); that is, growth of the whole population is geometric, until a high density is reached. Finally, the Allee type is shown in curve 3 (Figure 5-3); in this case growth rate per individual is at first directly density dependent and then inversely density dependent with optimum (for population growth rate) at an intermediate density. As with some previous diagrams in this book the curves in Figure 5-3 are intended to serve as general graphic models for the range of relationships to be found in nature. Intermediate situations would be expected, especially between patterns 1 and 2.

Among species with high reproductive potential, lemmings, which are small, mouselike rodents, are famous for "population irruptions" that occur every three or four years in the arctic. The lemming cycle resembles the curve shown in B-2, Figure 5-1. Another well-documented case of unstable populations involves needle-eating caterpillars (larvae of moths) in German pure-pine forests. Between 1880 and 1940 outbreaks of one or more species occurred every 5 to 10 years, when as many as 10,000 pupae per 1000 m² were present (producing enough caterpillars to defoliate the trees temporarily) as compared with the "normal" density of less than 1 in 1000 m². This case resembles A-1 in Figure 5-1. On the other hand, the size of most populations, even those with high reproductive potentials, remains remarkably constant year after year. As we have already pointed out, there is a general correlation between diversity and stability on the ecosystem level and homeostasis at the population level. That is irruptions and outbreaks are more likely to occur where the biological structure is simplified, either by man or by severe natural limiting factors, or when there is a sudden increase in energy or resources that gives the opportunistic species a chance to grow exponentially. Even in stable ecosystems, however, an occasional species will irrupt. One thing we can say for certain: We do not know enough about the details of population regulation to do a very good job of predicting when or where some species will break out of its usual limits.

COMPETITION The word "competition" denotes a striving for the same thing. At the ecological level competition becomes important when two organisms strive for something that is not in adequate supply for both of them. Thus, plants compete for light and nutrients in a forest, and animals compete for food and shelter when the latter are relatively scarce in terms of the density of animals. If the population consists of only a few scattered individuals, competition will not be a factor of ecological importance In the arctic, for example, plants may be so few and scattered due to the severe climate that no competition for light occurs.

The result of competition is that both parties (that is, the competitors) are hampered in some manner. At the population level this means that density, or rate of population energy flow, will be reduced or held in check by the competitive action. There is another type of interaction, known as *mutual inhibition,* that has the same result. This occurs when two organisms interfere with one another while striving for something even though that something is not in short supply. For example, the organisms might secrete substances that interfere with each other; or they might even eat each other. Many ecologists prefer not to include such direct mutual inhibition under the heading of competition. However, because we are interested in the results of interactions in terms of the community, we are perhaps justified in lumping all reciprocal negative interactions under the heading of competition.

Both intraspecific and interspecific competition can be very important in determining the kinds and numbers of organisms. Intraspecific competition is an important factor in those populations that tend to be self-regulated in the manner described above. An interesting behavior pattern, which results in intraspecific competition for space and a rather effective control of population size, is known as *territoriality*; it is characteristic of many species of birds and some other higher animals. At the beginning of the breeding season the male of a territorial species of bird will "stake out" a definite area of its habitat (that is, the "territory") and defend it against other males of the same species, with the result that no other male is allowed to enter the area. Much of the loud bird song we hear in the spring is for the purpose of announcing to other males "ownership" of land, and not for the purpose of wooing the female, as is often supposed. A male that is successful in holding his land has a high probability of mating and nesting, while a male that is unable to establish such a territory will not breed. Once the pair is formed, the female also joins in the defense of the territory in many species. The defended territory also serves a positive purpose of insuring the complex business of caring for the nest and young will not be interrupted by the presence of other males and females. Finally, it should be mentioned that the actual defense of the territory does not usually involve much actual fighting or other severe stress. Would-be invaders respect the established bird; loud songs or threat displays usually are sufficient to cause the invader to withdraw. However, if one of the pair should be killed it is very often quickly replaced by a bird from a reservoir of individuals not established. Therefore, since the territorial behavior pattern helps avoid both overcrowding and undercrowding it can be said to be regulatory in the sense that such behavior promotes a sigmoid growth pattern, or at least helps prevent an overshoot of carrying capacity.

Plant populations, as well as animal populations, may regulate themselves, so to speak, and avoid overcrowding. As we drive through the desert in the southwestern United States we are impressed with the fact that desert shrubs are widely spaced, often almost uniformly distributed as if planted at regularly spaced intervals. The pattern would seem to be explained by competition for scarce water, which eliminates all but one individual in a given area. However, evolution of self-regulation of the population has occurred in some species, in that severe competition for water is avoided by the production of leaf or root hormones that inhibit development of other individuals in the neighborhood. That is, chemical substances released by decaying leaves that fall to the ground under the shrub, or by the living roots in the soil, inhibit or kill any seedlings that may start to sprout. Such a control mechanism would tend to keep plants spaced apart, thereby reducing actual competition for water, which might result in stunting or death of all plants should there be many trying to grow in the same place. In this kind of "birth control" the quality of the individual in a limited environment is maximized, not the quantity; is there a lesson here for man?

COMPETITIVE EXCLUSION VERSUS COEXISTENCE

Where there are two or more closely related species adapted to the same or similar niche, interspecific competition becomes important. If the competition is severe, one of the species may be eliminated completely, or forced into another niche or another geographical location; or the species involved may be able to live together at reduced density by sharing the resources in some sort of equilibrium. These two possibilities are shown in Figure 5-4 in terms of growth form models based on two thoroughgoing experimental studies, one involving two closely related species of animals (beetles) and the other two species of plants (clovers).

Dr. Thomas Park at the University of Chicago, his students, and associates have carried out a long series of competition experiments with laboratory cultures of flour beetles, especially species belonging to the genus *Tribolium.* These small beetles can complete their entire life history in a very simple and homogeneous habitat, namely, a jar of flour or wheat bran. The medium in this case is both food and habitat for larvae and adults. If fresh medium is added at regular intervals, a population of beetles can be maintained for a very long time. In the energy-flow terminology discussed in Chapter 3 this experimental setup may be described as a stabilized heterotrophic ecosystem in which imports of food energy equal respiratory losses. (This laboratory ecosystem resembles the city or oyster reef, as diagrammed in Figure 2-2).

The investigators have found that when two different species of *Tribolium* are placed in this homogeneous little universe, invariably one species is eliminated sooner or later while the other one continues to thrive. In other words, one species always "wins" in the competition; or to put it another way, two species of *Tribolium* cannot both survive in the particular simple ecosystem, even though either one does quite well if alone in the culture. The relative number of individuals of each species originally placed into the culture does not affect the eventual outcome, but the "climate" imposed on the ecosystem does have a great effect on which species of a pair wins out. As shown in Figure 5-4, one species wins when the conditions are hot and wet, while the other species survives when conditions are cool

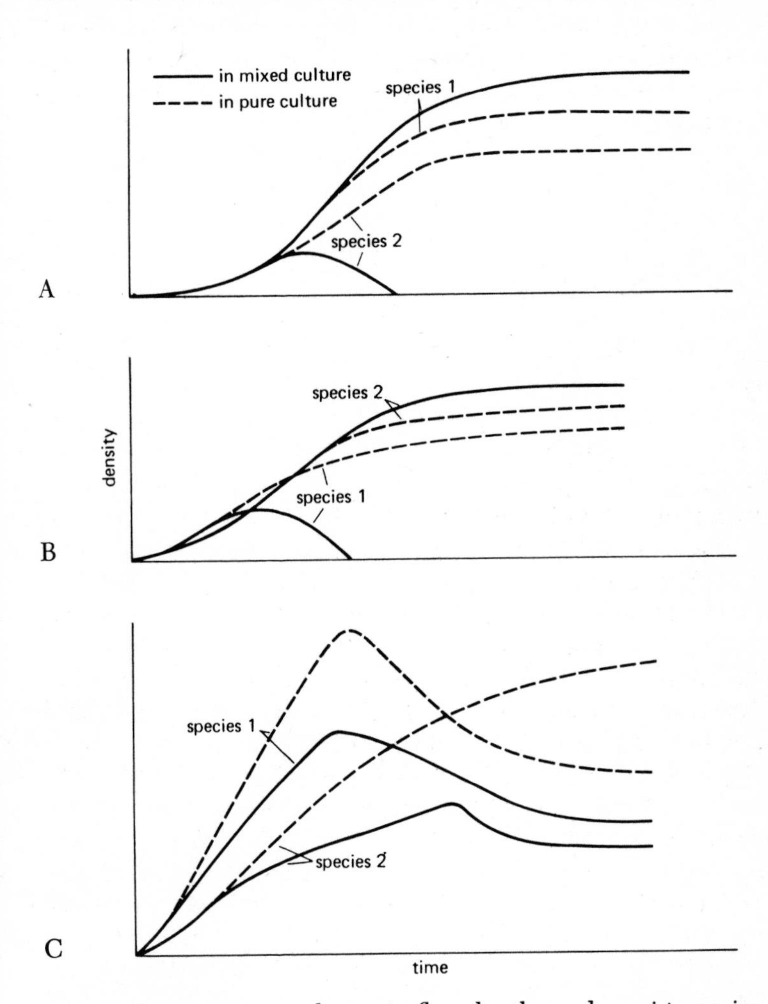

Fig. 5-4 Competition exclusion in flour beetles and coexistence in clovers. (A) Species of flour beetle 1, *Tribolium castaneum* excludes species; 2, *T. confusium* in mixed culture when "climate" is hot and wet (34°C, 70% R.H.) even though both species do well in pure culture. (B) Species 2, *T. confusium* excludes species 1, *T. castaneum* in mixed culture when "climate" is cool and dry (24°C, 30% R.H.) even though both species are also able to live separately under these conditions. (C) Two species of clover, *Trifolium repens* (species 1) and *T. fragiferium* (species 2) are able to coexist in mixed culture, but at reduced leaf density for each species as compared to pure cultures. (Graphs based on data of Park, 1954, for flour beetles and Harper and Chatsworthy, 1963, for clovers.)

and dry. Under intermediate conditions, sometimes one, sometimes the other species wins, with the percentage of wins and losses following the gradient between the extreme conditions. Halfway between the extremes the probability of either species winning might be 0.50, or a 50-50 chance. The elimination of one species by another as a result of interspecific competition has come to be known as the *competitive exclusion principle* (see Hardin, 1960) or Gause's principle, after an investigator who demonstrated such exclusion in cultures of protozoa in the 1930s.

That closely related species are able to coexist despite crowding and competition for limited resources is illustrated by experiments with clovers performed by Dr. J. L. Harper and associates at the University College of North Wales. The results on one series of experiments are shown in Figure 5-4C. Two species of clover of genus *Trifolium* were able to complete their life cycle and produce seed when thickly planted together in trays even though the density of each species was reduced compared with pure cultures. The solid lines in Figure 5-4C show population growth in mixed culture, while dotted lines are growth curves for the species in pure cultures. Small but important differences in morphology and population growth form enabled these two competitors to coexist. One species, T. *repens*, grows faster and reaches a peak in leaf density sooner, but the other species has longer peticles (leaf stems) and higher leaves and is thus able to overtop the faster growing species and thereby avoid being shaded out of existence. Harper (1961) concludes that two species of plants can persist together if there are differences in one or more of the following properties: (1) nutritional requirements (legumes and nonlegumes, for example); (2) causes of mortality (different sensitivity to grazing or toxins, for example); and (3) sensitivity to controlling factors (light, water, and so on) at different times, as in the case of the clover example. Much the same could be said for animals.

The study of laboratory or greenhouse populations contributes to our understanding of the mechanisms, ecological and genetic, that might operate when species interact. This is evident from the examples given, but, as emphasized a number of times, a multilevel approach is ultimately necessary since study at each level of organization contributes something, but not everything, to the total picture (see Chapter 1). When we shift our attention from the study of discrete populations under controlled conditions to the real world of communities and ecosystems we do indeed find good evidence for competition exclusion. However, more often we find that species, even those that cannot live together in a restricted microcosm, adapt to coexist by shifting their niches to reduce competition pressure through physiological and behavioral changes (that may or may not be reinforced by genetic selection), or by exploiting the same resources at different times (different seasons or time of day, for example), or by shifting their positions in environmental gradients. Perhaps we can illustrate some of these possibilities by an example.

Related species often replace one another rather abruptly in a natural gradient suggesting that interspecific competition might play a part in such a distribution. We could test this hypothesis by removing one species and observing whether the adjacent species invades the vacated area. One investigator (Connell, 1961) working with barnacles in an intertidal gradient in Scotland did just such an experiment. In Scotland, as well as along the seacoast of northeastern North America, two species of barnacles are commonly found on the rocky shores, one species of genus *Chthamalus* occupying a band near the upper part of the intertidal zone and another, larger species of genus *Balanus* occupying a wide band below the *Chthamalus* population as shown diagrammatically in Figure 5-5. Next time you take a walk along a rocky coast you can observe the zonal distribution of barnacles and other organisms for yourself. Although a zonal distribution might suggest competition, we could also explain it by assuming that physical factors in the gradient limit each species to a band quite independent

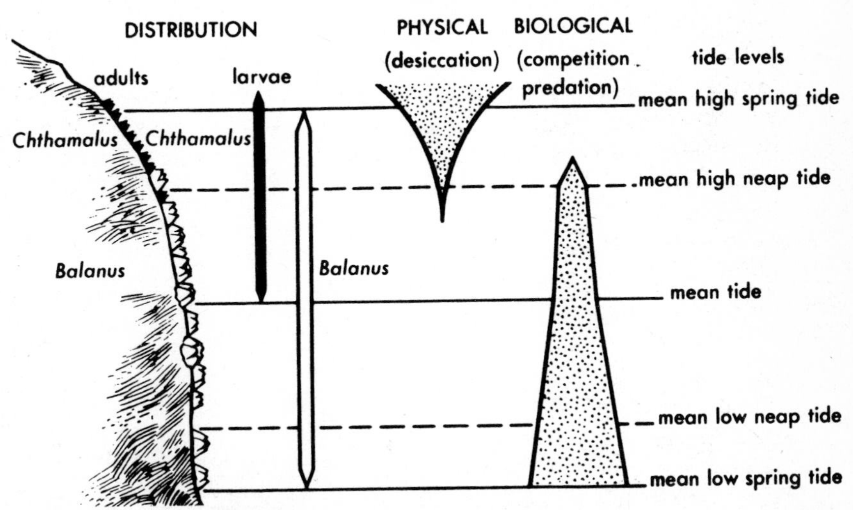

Fig. 5-5 Factors that control the distribution of two species of barnacles in an intertidal gradient. The young of each species settle over a wide range but survive to adulthood only within a more restricted range. Physical factors such as desiccation control upward limits of *Balanus* while biological factors such as competition and predation control downward distribution of *Chthamalus* in the lower portion of the intertidal gradient where physical environment is less limiting. (Redrawn from J. H. Connell, *Ecology*, Vol. 142, 1961.)

on the other. In the study in Scotland, it was found that larvae of both species tended to settle over a wider range than occupied by the adult population (Figure 5-5). When adult *Balanus* were cleared away and new ones kept from settling, the young *Chthamalus* survived and grew quite well in the upper part of the *Balanus zone* where they normally are not found. *Balanus*, however, did not extend into the *Chthamalus zone* even in the absence of *Chthamalus*. Figure 5-5 summarizes the study, which also included consideration of predators and the effect of desiccation. It was noteworthy that biological regulation, including inter- and intraspecific competition and predation, proved to be important in the middle and lower part of the gradient, but not in the upper section where exposure to drying and temperature extremes limited the population to one species and relatively few individuals. This model can be considered to apply to more extensive gradients such as an arctic-to-tropics or a high-to-low altitude gradient.

In another study, this one involving fiddler crabs in an intertidal marsh, experiments demonstrated that interspecific competition was only one of several factors keeping two species separated (Teal, 1958). In this case one species chose more often a muddy substrate and the other a sandy substrate; the sand-loving crab would invade the mud substrate if the other species was absent, but not too readily. It is probable that direct competition in the past played a part in the evolution of behavioral response to the substrate, which now plays the major role in keeping the species apart.

In addition to experimental evidence of the kind just cited there is a large body of observations that support the following generalization:

1. Closely related organisms often do not occur in the same place; or if they do, close study often shows that they use different energy sources, are active at different times of day or at different seasons, or otherwise occupy a different niche.
2. Where a large number of related species is present in a region, the niche of each is often narrower than when only a few species are present. Comparison of islands and mainlands often illustrates this trend. Thus, one investigator (Crowell, 1962) found that the cardinal was more abundant and occupied more marginal habitat in Bermuda, where the number of species of small birds (that is, potential competitors) was small, than on the United States mainland, where the cardinal is associated with a large number of other species.
3. Related species often replace one another in a gradient, as already mentioned.

In conclusion, we can say that competition together with other interactions between populations, to be discussed subsequently, are mechanisms that promote flexibility and a certain level of diversity in the community as a whole.

PREDATION Although the energy flow of predators (that is, secondary and tertiary consumers) is relatively small (see Figure 3-2), their role in regulating the primary consumers can be relatively great; in other words, a small number of predators can have a marked effect on the size of specific prey populations. On the other hand, as is also frequently the case, a predator may be only a minor factor in determining the size and growth rate of a prey population. As might be expected, there is a gradient of possibilities between these extreme possibilities:

1. The predator is strongly limiting to the point of reducing the prey to extinction or near extinction. In the latter case violent oscillations in the size of the prey population will result, and, if the predator cannot turn to other populations for food, violent oscillations in predator numbers will also occur.
2. The predator can be regulatory in that it helps keep the prey population from outrunning its resources, or, put another way, it contributes to the maintenance of a steady-state in the density of the prey.
3. The predator may be neither strongly limiting nor regulating.

Which situation exists for any pair of interacting species or groups of species depends on the *degree of vulnerability of the prey to the predator, as well as on the relative density levels and the energy flow from prey to predator.* From the predator's viewpoint this depends on how much energy it must expend to search for and capture the prey; from the prey's standpoint, this depends on how successfully individuals are able to avoid being eaten by the predator. A second principle about predator-prey interactions may be stated something as follows: *The limiting effects of predation tend to be reduced, and the regulating effects increased, where the interacting populations have had a common evolutionary history in a relatively stable ecosystem.* In other words, natural selection tends to reduce the detrimental effects of predation on both populations, since severe depression of a prey population by a predator can only lead to the extinction of one or both populations. Consequently, violent predator-prey interactions happen most frequently when the interaction is of recent origin (that is, when the two populations first become associated), or where there has been a recent large-scale disturbance in the ecosystem (as might be produced by man or by a climatic change).

Now that we have been so bold as to state two principles regarding predation, let us test them a bit and seek some examples. It is difficult for man to approach the subject of predation objectively. Although man himself is the greatest of all predators, often killing far beyond his needs, he tends to condemn all other predators without regard to the circumstance, especially if they prey on species he himself wishes to harvest. Sportsmen, among others, often have strong opinions against predators. The act of predation, such as a hawk catching a game bird, may be spectacular and easily observed whereas many other factors that may actually be more limiting to the bird population are not observed or are unknown to the untrained individual. For example, 30 years of objective study by Herbert L. Stoddard and his associates on the southwest Georgia game preserves have shown that hawks will not be a limiting factor to quail so long as vegetative escape cover lies near feeding areas so that birds can easily escape the attack of the hawk. Stoddard maintains high densities of quail by land management procedures that build up the food supply and refuge cover for the quail. In other words, his efforts are directed first toward improvement of the ecosystem specifically with the quail in mind. When this has been achieved, removal of hawks is unnecessary, even undesirable, because the quail are not vulnerable, and hawks also prey on rodents that eat quail eggs. Unfortunately, ecosystem management is more difficult and not so dramatic as shooting hawks. Game managers are often pressured into the latter even when they know better.

Now, let us cite an example of the opposite situation, where the predator exerts a marked effect. One of the author's students decided he could study a population of small rodents with more precision if he established a population on a small island in a new lake impoundment. Accordingly, he introduced a few pairs on the island, knowing that the animals could not leave it. For a while things went well; as the population grew, the student live-trapped and individually marked the animals and thus kept up with the births and deaths. Then, one day he went to the island to trap and found nothing. Investigation revealed a fresh mink den containing the bodies of many of his marked animals neatly cached away. Because the rodents were especially vulnerable on the island and could not escape or disperse, one mink had been able to find and kill them all.

To obtain an objective viewpoint it helps to think about predation from the population rather than from the individual standpoint. Predators, of course, are not beneficial to the individuals they kill, but they may be beneficial to the prey population as a whole. Some species of deer appear to be strongly regulated by predators. When the natural predators such as wolves, puma, bobcats, and so on, are exterminated, man has found it difficult to control deer populations even though by hunting he himself becomes the predator. In the eastern United States man at first overhunted and exterminated the native deer from large areas. Then, there followed a period of hunting restriction and of reintroduction, and deer again became abundant. Now deer are more abundant in many places than under primeval conditions, with the result that they are overgrazing their forest habitat and even dying from starvation during the winter. The "deer problem" has become especially acute in such states as Michigan and Pennsylvania, where large areas of second-growth forest provide maximum food favoring an almost geometric increase that is often not checked by the level of hunting. Two points can be made here: (1) Some predation is necessary and beneficial in a population that has been adapted to it (and that lacks self-regulation). (2) When man removes the natural control mechanism, he must replace it with an equally efficient regulatory mechanism if severe oscillations are to be avoided. An inflexible bag limit set without regard to the density, available food, and habitat has generally failed to bring about the desired regulation because such removal tends to be density independent. What is needed is density-dependent control (rate of removal increasing with density) since this is more regulatory, as noted earlier in this chapter.

Figure 5-6 shows a triangle of predatory interactions involving organisms that are not of direct economic importance to man; hence, we should be able to consider these data without bias. For a matter of years investigators at the University of Georgia Marine Institute at Sapelo Island have been studying the intertidal salt marsh as an ecosystem. The marsh is especially interesting to the ecologist because it is very productive yet contains only a limited number of species; hence, interrelations between populations are more easily studied. In the tall marsh grass growing along the salt creek banks live a small bird, the marsh wren, and a small rodent, the rice rat. Both feed on insects, snails, and, in the case of the rodent, small crabs and marsh grass. During the spring and summer the wren builds a globular nest out of grass in which to rear its young; during this season the rats prey on the eggs and young of the wren and often take over their nests for their own use. As shown in Figure 5-6, the energy flow between invertebrates and the two vertebrates is small in terms of the large populations of the former. Thus, the wren and the rat eat but a very small portion of their main food supply and, therefore, have little effect on the insect and crab populations; in this case predation is neither regulating nor controlling. In the annual cycle, wrens are a very minor food of the

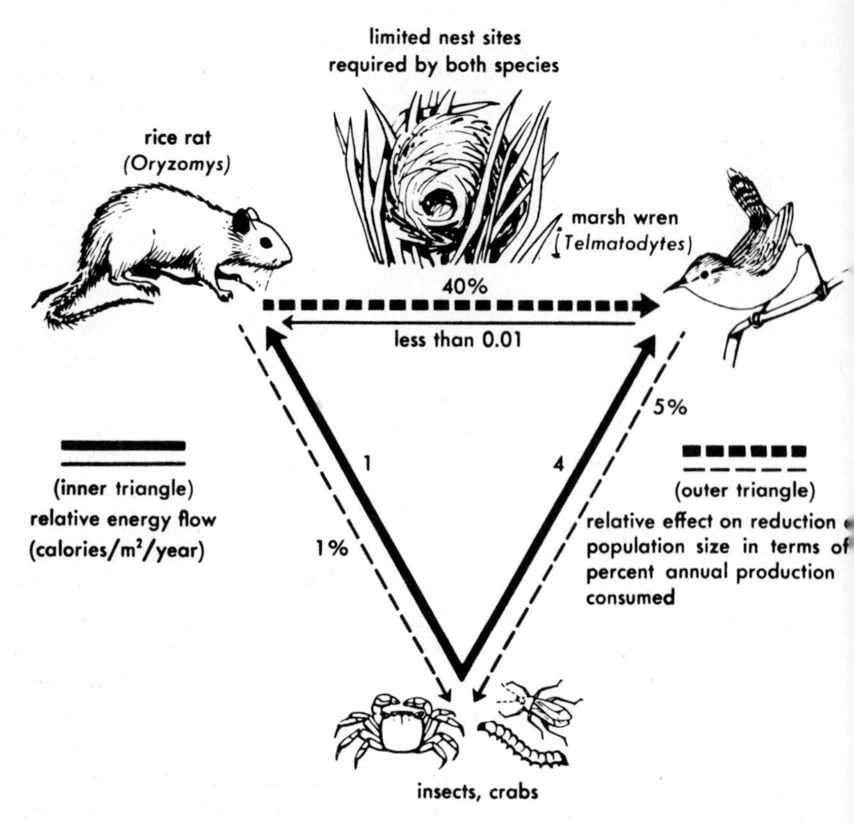

Fig. 5-6 A predator-prey triangle in a Georgia salt marsh ecosystem. The inner triangle shows estimated energy flow from prey to predator: the outer triangle is an estimate of the effect of predator on growth rate (production) of prey. Because the wren is vulnerable to the rice rat during the nesting season (when both species compete for nests in a limited habitat), the impact of predator on prey is not proportional to energy flow. See text for explanation. (Based on data of H. Sharp; H. Kale; and E. P. Odum.)

rat, yet because the wrens are especially vulnerable during the breeding season, the rat predation is a major factor in the wren mortality. When the density of rats was high, the wren population was depressed. Fortunately for the wrens, as yet undetermined factors limit the number of rats so that high density and resultant severe predation occur only in local areas.

We might consider the triangle between insects, rats, and wrens as a model for predation in general, since the triangular model shows how predation can be both a major and a minor item, depending on the relative density of predator and prey and the vulnerability of the prey to the predator. Remember, also, that the model does not mean that all bird-insect relationships are like this. The relationship depends on the species and the situation. Birds may be very effective predators on caterpillars that feed on leaf surfaces, and quite ineffective predators on leaf-mining insects that work inside the leaves. Likewise, insectivorous birds may exert effective control on insects when density of the latter is low (as is normally the case). But when an insect population irrupts (undergoes rapid exponential growth) birds may be ineffective because they cannot increase in numbers as fast as their prey.

PARASITISM Much of what has been said about predation also holds for parasitism. A population of parasites, whether they be bacteria, protozoans, fungi, helminth worms, or fleas, can be either strongly limiting, regulating, or relatively unimportant to a particular host population—in much the same manner as a predator may affect its prey. In fact, parasites and predators form a more or less continuous gradient ranging from tiny bacteria and viruses that live inside the host's tissues, to the large tigers of the ecosystem. Thet term *parasite* is usually used if the organism is small and actually lives in or on the host, which, therefore, is both energy source and habitat. In contrast, we think of a predator as being free living and larger than the prey, which serves as an energy source but not as habitat. All kinds of intermediate situations exist. The important point is that we are more interested in evaluating kinds of interactions than in classifying species into rigid categories. A given species may interact with other populations in more than one manner, depending on stage in the life cycle, locality, season, or other circumstance.

Although parasitism and predation are similar in terms of ecological regulation, important differences are found in the extremes of each situation. Parasitic organisms generally have higher reproductive rates and exhibit a greater "host" specificity than do most predators. Furthermore, they are often more specialized in structure, metabolism, and life history, as necessitated by their special internal environments and the problem of dispersal from one host to another. Some entire classes and orders or organisms, such as the Cestoda among the flatworms and the Sporozoa among the protozoans, have become adapted to parasitism. The most specialized species, such as the human malarial parasites, have a very complex life cycle that consists of a sequence of stages involving a succession of host tissues and an alternation of host species.

Host specificity of parasites is a very important consideration. Because many species of parasites can live only in one or a very few related species, the host-parasite interaction is especially intimate and potentially limiting to both populations. Man has often been able to utilize parasites in his attempt to regulate or control pests. In many cases insects that have been introduced from other parts of the world have been brought under control by transplanting the native parasite that regulated the insect in its original habitat; in other cases artificial propagation of parasites has helped. Practical biological control of this type is feasible with parasites that are specific for the species that man wishes to control. Such a parasite keeps constantly at work and can quickly adjust to increases and decreases in host numbers. In contrast, a pest can be rarely controlled by the introduction of a generalized predator that itself may become a pest if it spreads its attack to many other species. Thus, the English sparrow was introduced into New York City's Central Park for the purpose of controlling the elm spanworm. As we now know, this was a mistake; the sparrow not only did not control the insect (in the first place, tree caterpillars are not the main food item of this species) but it spread throughout the country and became a pest in many places. We have all heard about the introduction of the mongoose, a small, weasel-like predator, on islands; usually it was brought in for the purpose of controlling rats, but instead it often wreaks havoc with ground-nesting birds.

The cardinal principle concerning the relation between severity of interaction and length of association mentioned in the previou section is especially applicable to parasitism. In many cases the most severe diseases of man and the plants and animals on which he depends are caused by organisms so recently acquired, or recentl introduced, that the host has not developed protective immunity. The lesson for man, of course, is to avoid unnecessary introductions. Thus the citizen should be patient when he crosses state and national boundaries and is asked to avoid transporting biological material that might contain a potential pest or disease, We will come back to man's problems in disease and pest control in Chapter 8.

POSITIVE INTERACTIONS So far we have been concerned with what might be called the negative interaction between individuals or populations, that is, interactions that result in inhibition to one or both parties. Positive interactions are equally important in the ecosystem. The wide acceptance of Darwin's theory of "survival of the fittest" directed special attention to such factors a competition as a means of bringing about natural selection. However as Darwin himself pointed out, cooperation in nature is also important in natural selection. It seems reasonable to suppose that, like a balanced equation, negative and positive relations between popula tions eventually tend to balance if the ecosystem is to achieve any sort of stability.

We can think of positive interactions between populations of two or more species as taking three forms that, perhaps, represent an evolutionary series. *Commensalism* is a simple type of positive interaction in which one population benefits and the other is not affected to any measurable degree. A commensal relationship is especially common between sessile organisms and small motile organisms. The sea is a good place to observe such a relationship. Practically every worm burrow, shellfish, or sponge contains various "uninvited guests" that require the shelter or unused food of the host, but do neither harm nor good to the host. It is but a short step from this relation to parasitism on the one hand, or helpmate on the other. If two populations benefit each other, but are not essential to each other for survival, the relationship is often called *protocooperation*; if the association is necessary for the survival of both populations, we designate the interaction by the term *mutualism*.

The late W. C. Allee (1961) studied and wrote extensively on the subject of protocooperation. He believed that the beginning of cooperation between species can be seen throughout the animal and plant kingdoms. Cooperation is thus not restricted to the human population. In the sea, for example, crabs and coelenterates often associate with mutual benefit even though the two species can live separately; the coelenterates grow on the backs of crabs, where they provide camoflage and protection (since these animals have stinging cells) for the crab. In return, the coelenterate is transported about and obtains particles of food when the crab captures and eats other animals.

Mutualism is extremely widespread and important. The mutualistic association of legumes and nitrogen-fixing bacteria, and also mycorrhiza and trees, has already been mentioned in the preceding chapter. Mutualism often involves species of very different taxonomic relationships. One whole group of plants, the lichens, are examples; in this case, algae and fungi are so closely associated that the botanist finds it convenient to consider the association as a single species. The lichen, in a sense, is a tiny ecosystem containing autotrophic (algae) and heterotrophic (fungi) components.

It is probably that protocooperation and mutualism can develop or evolve not only from commensalism but also from parasitism; as already hinted, parasites and hosts may become actually beneficial (from the population standpoint) after a long association. In some primitive lichens, for example, the fungi actually penetrate the algal cells, as shown in Figure 5-7, and are thus essentially parasites of the algae. In the more advanced species the fungi mycelia do not break into the algal cells, but the two live in close harmony (Figure 5-7). So successful is partnership that lichens are able to live in the harshest physical environments such as granite outcrops and arctic tundras.

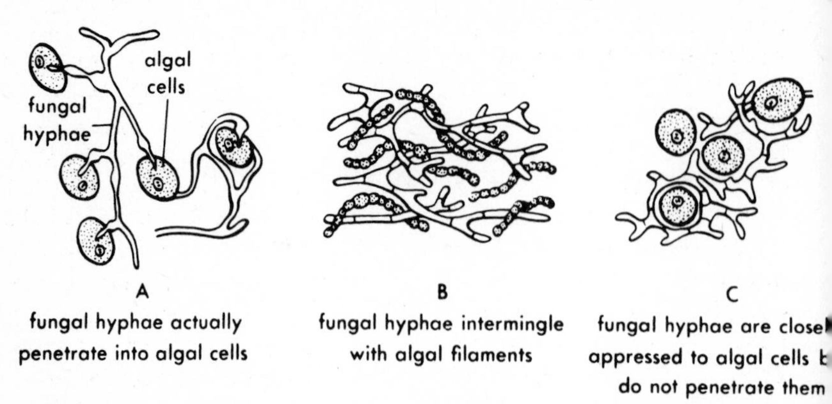

Fig. 5-7 A trend in evolution from parasitism to mutualism in the lichens. In some primitive lichens the fungi actually penetrate the algae cells, as in diagram (A), whereas in more advanced species the two organisms live in greater harmony for mutual benefit, as in (B) and (C).

In one sense mutualism is a model for a regulated ecosystem where even parasites and predators are useful in the sense that they "cooperate" for mutual survival. Man has made considerable progress in cooperation within his own population, and he is now also turning greater attention to achieving greater cooperation with other organisms for mutual benefit. There is much to be gained if we can transform negative interactions into positive ones. Again we come back to the theme with which this book began; that is, man thrives best when he functions as a part of nature rather than as a separate unit that strives only to exploit nature for his immediate needs or temporary gain (as might a newly acquired parasite). Since man is a dependent heterotroph, he must learn to live in mutualism with nature; otherwise, like the "unwise" parasite, he may so exploit his "host" that he destroys himself.

NUTRIENT CYCLING AND ENERGY FLOW AT THE POPULATION LEVEL

Although the emphasis in Chapters 3 and 4 was on energy flow and nutrient cycling in large ecosystems it is equally worthwhile to study these basic processes at the population level. An example is the work of Kuenzler (1961), who has measured both energy flow and the cycling of phosphorus in a population of shellfish living in a salt-marsh ecosystem. The ribbed mussel, *Modiolus demissus*, is one of the abundant animals living in salt marshes that form a belt between the outer barrier islands and the mainland along the coast of the southeastern United States. The mussel, which is about the size of an oyster, lives in small colonies attached to the substrate throughout the marsh. On the basis of random samples, Kuenzler found that the numbers varied from about $8/m^2$ average to $32/m^2$ in the more favorable spots. Organic biomass (ash-free dry weight) averaged 12 g, (equivalent to 60 $kcal/m^2$). Measurements of growth and respiration indicated that the population energy flow was a modest 56 $kcal/m^2/year$. However, in obtaining its food of small organisms and particles from the seawater that covers the marsh with each tide, the mussels filter very large quantities of water. As a result, a large amount of particulate matter or detritus is removed from the water and sedimented on the surface of the marsh. The particles, which are rich in phosphorus, other minerals, and vitamins, are thus retained in the marsh by action

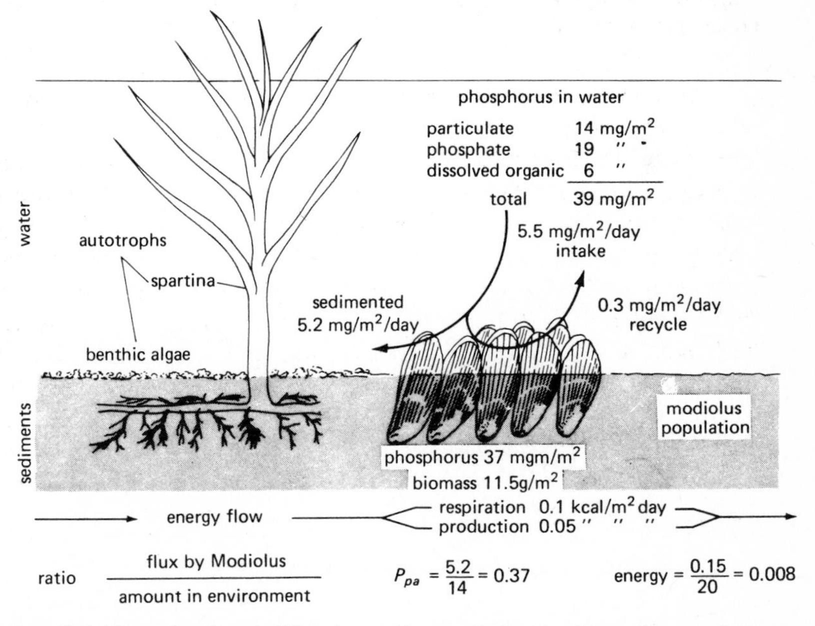

Fig. 5-8 The role of a shellfish (mussel) population in the cycling and retention of phosphorus in an estuarine ecosystem. The population has a major effect on the distribution of phosphorus even though the species is but a small component in the community in terms of biomass and energy flow. (Based on data from E. J. Kuenzler, *Limnology and Oceanography*, Vol. 6, 1961.)

of the mussel instead of being carried out by the tide. Kuenzler was able to calculate that the turnover time of particulate phosphorus in the water due to the action of the mussel population was only 2.6 days That is, the mussels removed from the water every 2½ days a quantity of phosphorus equal to the average amount present in the particles in the water (estimated to equal about 14 mg/m^2). The population was also found to have marked, although less, effect to cycling of dissolved phosphorus. The study is summarized in Figure 5-8, which shows amounts and flux rates.

Although the mussels are a relatively minor component of the marsh in terms of biomass and energy flow, they proved to have a major effect on the cycling and retention of valuable phosphorus. The mussel is not particularly important as a direct source of food for man or other animals, since the production (or growth) per year is not great in comparison with production of other populations, but the species is important as an agent helping to maintain fertility and, thereby, production of autotrophs. To put it another way, Kuenzler's study demonstrated that the mussel population is more important to the ecosystem as a biogeochemical agent than as a transformer of energy. Here again is an excellent illustration of the point previously made; species in nature may have great value to man in an indirect way not apparent on superficial examination. A species does not have to be a link to man's food chain to be valuable. Man needs the help of many species in maintaining stability and fertility of his environment.

MATHEMATICAL MODELS OF POPULATIONS

Because birthrates, death rates, and growth rates can be so precisely and neatly expressed by differential equations, mathematical models of population phenomena are well advanced. Starting with the pioneer work of Lotka and Volterra in the 1920s mathematical models of predator-prey and parasite-host interactions have been special favorites with ecologists, and such models have provided a useful basis for predicting pest and disease outbreaks. An introduction to this kind of mathematics can be found in most any upper undergraduate college level textbook of ecology, as well as in many special books and review papers. An important school of systems ecologists (that is, ecological modellers) started with such two-species models which are now being extended to the community and ecosystem level (for a nontechnical review see Holling, 1973).

HUMAN POPULATION GROWTH

The human population has experienced just about every kind of growth form imaginable, including negative growth, such as during the fourteenth century when the bubonic plague (black death) reduced the population of Europe about 25 percent. For many centuries the human population grew very slowly, if at all, and then came two periods of more rapid increase directly tied to energy procurement. The first major increase came with the development of agriculture (that is, the subsidized solar-powered system) beginning about 8000 years ago, and the second and much more rapid increase started about 200 years ago with the industrial revolution, the development of the fuel-powered system, and a decrease in death rates resulting from advances in medicine and health care. About 80 percent of the increase in human numbers since the dawn of man has occurred during the past two centuries. Some anthropologists take the viewpoint that agricultural and industrial development is as much a result of, as a cause of, this recent rapid population growth (see Spooner, 1972).

A good way to visualize human population growth is in terms of an annual rate of increase, which we call r and the doubling time (that is, number of years required to double the density) which we call t. Annual growth is generally expressed in terms of the number of persons added per thousand, or per hundred expressed as a percentage. Thus, a growth rate of 2 percent per annum means that 2 persons per 100 or 20 per 1000 are added each year. Since over the years people added to the population themselves produce more people, growth at 2 percent is more than just adding 2 per 100 each year; the population actually tends to increase in compound interest fashion. An approximate estimate of doubling time for a 2 percent rate can be obtained as follows:

$$t = \frac{\log^{e} 2}{r}\ ; t = \frac{0.6931}{0.02} = 35 \text{ years}$$

Two percent does not seem like much, but at that rate the population of your city could double twice in your lifetime. Through most of human history the growth rate averaged less than 0.1 percent per year (doubling time 700 or more years). At the present time world population growth is about 2 percent. The population of many undeveloped countries (which comprise better than two thirds of the world's total population) is still growing at a 3 percent rate (doubling time about 23 years), but in the developed countries the rate is dropping below 1 percent (doubling time 70 years or more). However, urban growth continues in the latter countries at better than 3 percent. Thus, the social disorder inherent in crowding and mismatched development rates can be severe even in countries where the overall growth rate approaches zero. For more on the history of the human population see Ehrlich and Ehrlich (1972, Chapters 2 and 3); Freedman and Berelson (1974); and Coale (1974).

The basic principles of population growth and its regulation, as discussed in this chapter, are relevant to the human population situation. However, man differs from other organisms not only in his relatively greater power to control, but in his development of a complex culture that differs widely in different parts of the earth. Usually, culture is adaptive, but not infrequently cultural patterns of behavior persist long after they cease to be advantageous. A high birthrate, for example, is an advantage when density is low and resources unused but disadvantageous when density is high and/or resources limited. In past periods of rapid population increase the death rate drops first while the birthrate remains high. If the standard of living rises then the birthrate declines sharply, thus reducing the rate of increase. For this reason, most political leaders (especially in the undeveloped countries) and some population researchers as well believe that raising the standard of living is the primary route to population control. However, if energy and materials are limiting, as is now evident in large areas of the globe, rapid population growth becomes a hindrance rather than a help to economic and other development that might raise the standard of living for masses of people.

Suffice it to say that most students of the human population feel that in addition to efforts to raise the standard of living some more direct forms of self-regulation are now necessary—as, for example—positive programs for birth control, tax incentives for reduced family size, women's rights movements, educational programs to change public attitudes, and so on. As we have noted several times in this book there is great danger in the instability and disorder that is generated when large, high energy-consuming populations overshoot one or more supporting functions. Because of the time lags mentioned on page 124 exponential human population growth in some parts of the world at least, seems headed for just such an overshoot unless regulatory efforts can be undertaken well before a critical limit is overshot. Thus, the alternatives to self-regulation, such as pestilence, starvation, severe pollution, war, and social upheavals that might reduce the unmanageable rate of increase are not pleasant to ponder. The choice seems to be either reduce the birthrate or suffer an increase in the death rate. The productivity and carrying capacity of the biosphere for man can and must be increased to accommodate the several more doublings that we are already committed to regardless of whether we take any action or not. But the sooner we reduce the rate of increase (that is, "cool" the "boom" or "inflation" in human numbers) the better we will be able to cope with the myriad problems created by the first (and we hope the last) population explosion ever experienced by mankind.

It is important to note that the optimum population density for a highly developed, industrialized nation with a high per capita energy consumption may be quite a bit lower than the population that can be supported at a subsistence level when we consider the whole of the ecosystem. The confusing relationship here is that energy-intensive industrial development increases the number of people that can live crowded together in urban and suburban areas, but many acres of relatively vacant land and water and huge amounts of natural resources are required to support these "hot spots" (see Chapter 2). The problem of determining "carrying capacity" will be discussed in Chapter 8. What may be a bit of good news is that early in 1974 the United Nations issued a revised forecast that the world population density would level off after about two more doublings; this forecast is quite different from the one made earlier to the effect that the world population would increase beyond this level.

SUGGESTED READINGS

References cited

Allee, W. C. 1961. *Cooperation Among Animals with Human Implications.* New York: Henry Schuman. (See also paperback edition: *Social Life of Animals.* Boston: Beacon Press. 1958.)

Coale, Ansley J. 1974. The history of the human population. *Sci. Amer.* 231(3):40–51.

Connell, Joseph H. 1961. The influence of interspecific competition and other factors on the distribution of the barnacle, *Chthamalus stellatus. Ecol.* 42:133–146.

Crowell, K. L. 1962. Reduced interspecific competition among the birds of Bermuda. *Ecol.* 43:75–88.

Ehrlich, Paul R. and A. H. Ehrlich. 1972. *Population, Resources and Environment,* 2nd ed. San Francisco: W. H. Freeman.

Freedmann, R. and B. Berelson. 1974. The human population. *Sci. Amer.* 231 (3):30–39.

Hardin, G. The competitive exclusion principle. *Science.* 131:1292–1297.

Harper, John L. 1961. Approaches to the study of plant competition. In Mechanisms in Biological Competition. *Symposium Soc. Exp. Biol. Number XV.*, pp. 1–268. England: Cambridge Univ. Press.

Harper, J. L. and J. N. Clatworthy. 1963. The comparative biology of closely related species VI analysis of the growth of *Trifolium repens* and *T.*

fragifesum in pure and mixed populations. *J. Exp. Bot.* 14:172–190.

Holling, C. S. 1973. Resilience and stability of ecological systems. *Ann. Rev Ecol. and System.* 4:1–21.

Kuenzler, Edward J. 1961. Phosphorus budget of a mussel population *Limnol. and Oceanogr.* 6:400–415.

Park, Thomas. 1954. Experimental studies on interspecific competition *Physiol. Zool.* 27:177–238.

Spooner, Brian, ed. 1972. *Population Growth. Anthropological Implications* Cambridge, Massachusetts: M.I.T. Press. (Several authors discuss the general concept that human population growth is the "forcing function" that brings on intensive agriculture and centralized government.)

Teal, John M. 1958. Distribution of fiddler crabs in Georgia salt marshes *Ecol.* 39:185–193.

Population dynamics

MacArthur, Robert and Joseph Connell. 1966. *The Biology of Populations* New York: John Wiley & Sons.

Odum, E. P. 1971. *Fundamentals of Ecology,* 3rd ed. Chapters 7 and 8 Philadelphia: Saunders.

Solomon, M. E. 1969. Population dynamics. *Studies in Biology,* No. 18 New York: Edward Arnold, London and St. Martins Press.
Sinauer Associations: Stamford, Connecticut.

Wilson, E. O. and W. H. Bossert. 1971. *A Primer of Population Biology*

Wynne-Edwards, V. C. 1964. Population Control in Animals. *Sci. Amer* 211(2):68–72.

Population interactions (competition, mutualism, and so on)

Ayala, Francisco J. 1972. Competition between species. *Amer. Sci.* 60(3): 348–357.

Burkholder, Paul R. 1952. Cooperation and conflict among primitive organisms. *Amer. Sci.* 40:601–631. (Classifies nine types of interspecific interactions on basis of positive and negative effects.)

Henry, S. M., ed. 1966. *Symbiosis.* New York: Academic Press.

Howard, Walter E. 1974. The biology of predator control. *Addison-Wesley Module in Biology,* No. 11. Reading, Massachusetts: Addison-Wesley.

Miller, Richard S. 1967. Pattern and process in competition. *Adv. in Ecol. Res.* 4:1–74. New York: Academic Press.

Milthorpe, F. L., ed. 1961. *Mechanisms in Biological Competition.* Cambridge Univ. Press.

Nutman, P. S. and B. Mosse, eds. 1963. Symbiotic associations. *Thirteenth Symposium of Society for General Microbiology.* Cambridge Univ. Press.

Park, T. 1962. Beetles, competition and populations. *Science,* 138:1369–1375.

Pimentel, David. 1968. Population regulation and genetic feedback. *Science.* 159:1432–1437. (Evolutionary tendency for severe negative interspecific effects to be reduced or to become positive.)

_______. 1973. Genetics and ecology of population control. *Addison-Wesley Module in Biology*, No. 10. Reading, Massachusetts: Addison-Wesley.

The human population

Borgstrom, Georg. 1969. *Too Many: A Study of the Earth's Biological Limitations.* New York: MacMillan.

Calhoun, J. B. 1962. Population density and social pathology. *Sci. Amer.* 206:1399–1408.

Ehrlich, Paul R. 1968. *The Population Bomb.* New York: Ballantine Books.

Enke, Stephen. 1969. Birth control for economic development. *Science.* 164:798–802. (Reducing human fertility can raise per capita income in less-developed countries.)

Galle, Omer R.; W. R. Gove; and J. M. McPherson. 1972. Population density and pathology: what are the relations for man? *Science.* 176:23–30. (Evidence from one city suggests that high population density may be linked with "pathological" behavior, as found in Calhoun's animal experiments, see reference in this list.)

Holdren, John P. and P. R. Ehrlich. 1974. Human population and the global environment. *Amer. Sci.* 62:282–292.

Keyfitz, Nathan. 1971. On the momentum of population growth. *Demography*, 8:71–80. (The tendency for rapid growth to "overshoot.")

National Academy of Science. 1971. *Rapid Population Growth: Consequences and Policy Implications.* Prepared by a Study Committee, Roger Revelle, Chairman. Baltimore: Johns Hopkins Press. (Conclusion: rapid population growth has more disadvantages than economic advantages.)

Sci. Amer.: Special Issue, on the Human Population. September 1974, Vol. 231, No. 3. (Eleven well-prepared articles dealing with historical, social, political, and biological aspects, and the contrasting situation in developed and undeveloped countries.)

Selye, Hans. 1973. The evolution of the stress concept. *Amer. Sci.* 61:692–699. (Stress as the nonspecific response of the body to demands made on it is an important medical concept in relation to population pressure and toxic substances in the environment.)

See also, suggested readings for Chapter 8.

chapter 6

Ecosystem Development and Evolution

One of the most dramatic and important consequences of biological regulation in the community as a whole is the phenomenon generally known as ecological succession, but better described by the phrase, *ecosystem development*. When a cultivated field is abandoned in the eastern part of North America, for example, the forest that originally occupied the site returns only after a series of temporary communities have preceded it (see Figure 6-2). The successive stages may be entirely different in structure and function from the forest that eventually develops on the site. In fact, we may think of such temporary communities as developmental stages analogous to the life-history stages through which many organisms pass before reaching adulthood. Capacity for self-development constitutes an important property that distinguishes systems with major biological components from systems that are primarily physical. Models of ecological systems that fail to include short-term developmental and longer-term evolutionary processes will fall short of the mark. In other words, *when dealing with ecosystems we must include developmental parameters*[1] *in addition to parameters derived from physical laws* (such as laws of thermodynamics).

To look at the situation in another way we can say that change with time in ecosystem structure and function results from an interaction of physical forces impinging from without (recall the discussion of the concept of "forcing function" in Chapter 1) and developmental processes generated within the system. For convenience we may speak of a sequence of changes primarily due to the former as *allogenic succession* (allo = outside, genic = relating to), and internally generated sequences as *autogenic succession* (auto = self-propelling) or *autogenic development.* As we shall see, allogenic processes dominate some ecosystems and autogenic processes others. But first let us consider autogenic development as a unique feature of most ecosystems.

DEFINITIONS Ecosystem development as an autogenic process may be defined in terms of the following three parameters: (1) It is the orderly process of community changes; these are directional and, therefore, predictable. (2) It results from the modification of the physical environment and population structure by the community. (3) It culminates the establishment of as stable an ecosystem as is biologically possible on the site in question. It is important to emphasize that this kind of ecological change is *community controlled*; each set of organisms changes the physical substrate and the microclimate (local conditions of temperature, light, and so on), and species composition and diversity is altered as a result of competitive and other population interactions described in Chapter 5. When the site and the community has been modified as much as it can be by biological processes, a steady-state develops—at least in theory. Also, in theory energy utilization is optimized in that maximum biomass (or information content) is maintained per unit of available energy flow. The species involved, time required, and degree of stability achieved depend on geography, climate, substrate, and other physical factors, but the process of development itself is biological, not physical. That is, the physical environment determines the pattern of change but does not cause it.

In summary, increasing the efficiency of energy utilization so that each unit of structure is maintained with the least possible work can

[1] The term parameter refers to a characteristic element or constant factor; in terms of a model it is a factor that can be assumed without necessity of measurement or direct observation.

be considered to be "the strategy of ecosystem development." Strong physical forces or surges, as well as large harvests or pollution input from man's fuel-powered systems, will modify, halt, or abort this developmental course. As we shall see, understanding man's impact on the developmental process is one of the most important considerations in achieving a reasonable working balance between man and nature.

SOME BASIC TERMS In ecological terminology the developmental stages are known as *seral stages,* and the final steady-state as the *climax.* The entire gradient of communities that is characteristic of a given site is called a *sere.* Succession that begins on a sterile area where conditions of existence are not at first favorable—as, for example, a newly exposed sand dune or a recent lava flow—is termed *primary succession.* The term *secondary succession* refers to community development on sites previously occupied by well-developed communities, or succession on sites where nutrients and conditions of existence are already favorable, such as abandoned croplands, plowed grasslands, cut-over forests, or new ponds. As would be expected, the rate of change is much more rapid, and the time required for the completion of the sere is much shorter, in secondary succession. Finally, it is important to distinguish between what may be called (for lack of better terms) *autotrophic succession* and *heterotrophic succession.* The former is the widespread type in nature that begins in a predominantly inorganic environment and is characterized by early and continued dominance by autotrophic organisms. Heterotrophic succession characterized by early dominance by heterotrophs occurs in the special case where the environment is predominantly organic as, for example, in a stream heavily polluted with sewage or, on a smaller scale, in a fallen log. Energy is maximum at the beginning and declines as succession occurs unless additional organic matter is imported or until an autotrophic regime takes over. In contrast, energy flow does not necessarily decline in the autotrophic type but is usually maintained or increased during succession.

LAB CULTURE MODELS OF SUCCESSION If glass flasks half-filled with a culture media containing a good balance of inorganic salts necessary for life are inoculated with samples of water and sediments from a pond, ecological development on a microcosm scale is set in motion if the flasks are placed under a good light source as in a laboratory growth chamber.

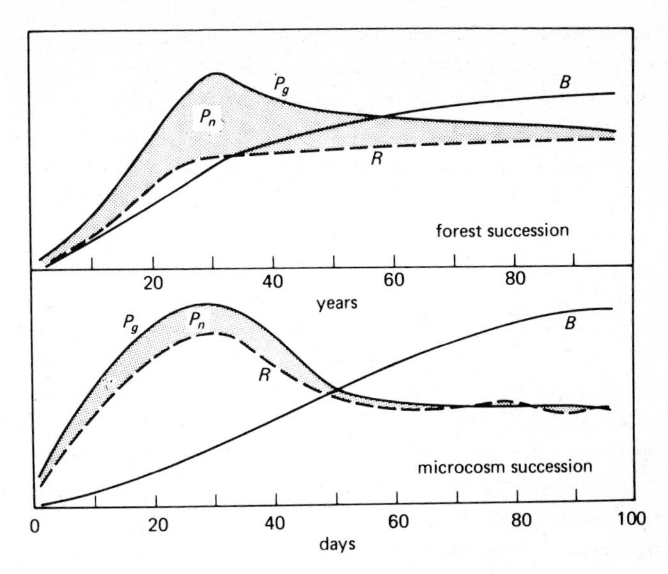

Fig. 6-1 Comparison of the energetics of ecosystem development in a forest (redrawn from Kira and Sludei, 1967) and a laboratory microcosm (redrawn from Cooke, 1967). P_g = gross primary production; P_n = net primary production; R = total community respiration; B = total biomass. (After E. P. Odum, *Fundamentals of Ecology,* 3rd ed. W. B. Saunders, 1971, p. 254.

It is a good practice to cross-inoculate the flasks for the first few days to make sure that a variety of small organisms or their propaguiles (both plant and animal) get into each container. Figure 6-1, lower diagram, shows how three properties of the ecosystem, namely production by photosynthesis (P), respiration (R), and biomass (B) change as the microcosm community undergoes succession. The basic pattern of change parallels that which occurs in the longer, open-system development of a forest, as shown on the upper diagram of Figure 6-1. During the first few weeks the autotrophs (algal cells) introduced with the pond water exploit the temporarily unlimited nutrients and undergo exponential growth (see page 125). Small heterotrophs (bacteria, protozoa, nematodes, crustaceans, and so on) respond likewise, so that total volume of living material or biomass increases rapidly. During this youthful growth stage the total or gross production rate (P_G) exceeds total respiration (R) so that $P/R > 1$. As resources (space, nutrients, and so on) reach saturation use in the closed system the rate of production becomes limited by rate of decomposition and regeneration of nutrients. A climax steady-state then develops in which production balances respiration, $P = R$, and $P/R = 1$. In this stage there is no net production (P_N), and no further net increase in biomass in the system. Other characteristics of the climax include a high ratio of biomass to daily production rate which means slow turnover (see page 28 for discussion of the concept of turnover), and a large amount of organic material (B) maintained by a relatively small increment of daily energy expenditure (R). We will discuss the significance of this kind of efficiency later in the chapter. In appearance, the cultures go from a bright green color to a yellow-green color with detritus and detritus-consuming animals becoming more prominent in the mature stage. Climax cultures may continue indefinitely, but if the diversity in the original inoculation was low they may gradually age and die. New successions can be set in motion at any time by inoculating from old cultures into new media, or by adding new media to old cultures.

As indicated in the previous section, succession may proceed either from the extremely autotrophic condition where P exceeds R, as in the cultures we have just described, or from the extremely heterotrophic condition where R exceeds P (or where P may be zero). An interesting culture model of the heterotrophic type of succession is the familiar hay infusion often used for growing protozoans and other small animals for students to study in the elementary biology laboratory. If a quantity of dried hay is boiled, and the solution allowed to stand a few days, a thriving culture of heterotrophic bacteria develops. If some pond water containing seed stocks of various small animals is then added to the hay infusion, a succession of species can be observed for about a month. Usually, small flagellates called monads appear first, followed in rapid succession by ciliated protozoans such as Colpoda and Paramecium; changes then come more slowly, with specialized ciliates (such as Hypotricha and Vorticella), Amoeba, or rotifers reaching peaks of abundance. If algae get into the culture, then an equilibrium with P nearing R may be approached; otherwise, the culture will run down in about 90 days since all the organism will die for lack of food, the original organic matter introduced having been used up.

Thus, the two types of succession can be contrasted on a small scale in the laboratory, or as a class exercise in a course in ecology, by an algae culture on inorganic media and a hay infusion, as a culture starting with an organic medium. Such cultures demonstrate, respectively, what happens in the early stages of succession in a new pond or artificial lake, and in the early stages of succession following dumping of sewage or other organic wastes into a pond or stream. For more on laboratory "microecosystems" or "microcosms" as models for ecological succession, see Beyers (1964) and Cooke (1967). In general, laboratory cultures are much too small and too closed, and do not contain enough diversity (biological or physical) to reveal all the important features of ecosystem development.

A TABULAR MODEL FOR ECOSYSTEM DEVELOPMENT

A more general and complete summary of important changes in community structure and function in the sere, as revealed by the study of the large, open systems of nature, is shown in Table 6-1.

Expected trends in the gradient from youth to maturity are grouped under several headings. Although ecologists have studied succession in many parts of the world, most of the emphasis to date has been on the descriptive aspects such as the qualitative changes in species structure. Only recently have the functional aspects of succession also been considered. Consequently, some of the items listed in Table 6-1 must be considered hypothetical in the sense that they are based on good experimental or theoretical evidence, but have not been verified by adequate data from the field. Five aspects seem most significant and require a bit more explanation as follows:

The kinds of plants and animals that change continuously with succession. Those species that are important in the pioneer stages are not likely to be important in the climax. When the density of species in a sere is plotted against time, a characteristic stair-step graph is obtained, as illustrated in Figure 6-2. Such a pattern usually is apparent whether we are considering a specific taxonomic group, such as birds, or a trophic group, such as herbivores or producers. Typically, some species in the gradient have wider tolerances or niche preferences than others and, therefore, persist over a longer period of time. Thus, in the terrestrial succession pictured in Figure 6-2 pine trees and cardinals persist through longer periods of time than do most of the other species. In general, the more species in the group (whether taxonomic or ecological) that are geographically available for colonization, the more restricted will be the occurrence of each species in the time sequence. This kind of regulatory adjustment is the result of competition-coexistence interactions discussed in the preceding chapter.

Biomass and the standing crop of organic matter increase with succession. In both aquatic and terrestrial environments the total amount of living matter (biomass) and decomposing organic materials (detritus and humus, see Chapter 2) tend to increase with time. Also, many soluble substances accumulate; these include sugars, amino acids, and many organic products of microbial decomposition. These liquid products that leak out from the bodies of organisms are often collectively known as *extrametabolites.* Some of these substances provide food for microorganisms, and perhaps also for macroorganisms. Other substances are equally important in that they may act as inhibitors (antibiotics) or as growth promoters (as, for example, vitamins). Substances produced by one organism may inhibit the further growth of that species (thus providing population self-regulation, see page 124)

Table 6-1. A Tabular Model for Ecological Succession of the Autogenic, Autotrophic Type

Ecosystem characteristic	Trend in ecological development early stage to climax or youth to maturity or growth stage to steady-state
Community Structure	
Species composition	changes rapidly at first, then more gradually
Size of individuals	tends to increase
Number species of autotrophs	increases in primary and often early in secondary succession; may decline in older stages as size of individuals increases
Number species of heterotrophs	increases until relatively late in the sere
Species diversity	increases initially, then becomes stabilized or declines in older stages as size of individual increases
Total biomass	increases
Nonliving organic matter	increases
Energy Flow (Community Metabolism)	
Gross production (*P*)	increases during early phase of primary succession; little or no increase during secondary succession
Net community production (yield)	decreases
Community respiration (*R*)	increases
P/R ratio	$P > R$ to $P = R$
P/B ratio	decreases
B/P and B/R ratios (biomass supported/unit energy)	increases
Food-chains	from linear chains to more complex food webs
Biogeochemical Cycles	
Mineral cycles	become more closed
Turnover time	increases
Role of detritus	increases
Nutrient conservation	increases
Natural Selection and Regulation	
Growth form	from *r*-selection (rapid growth) to *K*-selection (feedback control)[a]
Quality of biotic components	increases
Niches	increasing specialization
Life cycles	length and complexity increases
Symbiosis (living together)	increasingly mutualistic
Entropy	decreases
Information[b]	increases
Overall stability	increases

[a] See text for explanation.

or they may act on completely different species. This was dramatically brought to our attention by the discovery of penicillin and other bacterial antibiotics produced by fungi. In other cases, increasing organic matter stimulates the growth of bacteria that manufacture vitamin B_{12}, a necessary growth promoter for many animals (many are unable to manufacture this and other vitamins themselves). Where extrametabolites do prove to be regulatory, we would be justified in calling these substances *environmental hormones* since by definition a hormone is a "chemical regulator." Chemical regulation is one way of achieving community stability as the climax is approached, because the physical as well as the chemical perturbations (as, for example, light and water relations) are buffered by a large organic structure. There is no question that *the increase in amount of and the change in organic structure are two of the main factors bringing about the change in species during ecological development.*

The diversity of species tends to increase with succession. Initially this is the case, although it is not clear from the present data that the change in variety of taxa follows the same pattern in all ecosystems. Increase in diversity of heterotrophs is especially striking; the variety of microorganisms and heterotrophic plants and animals is likely to be much greater in the later stages of succession than in the early stages. Maximum diversity of autotrophs in many ecosystems seems to be reached earlier in succession. The interplay of opposite trends makes it difficult to generalize in regard to diversity. The increase in size of individual organisms and the increase in competition tend to reduce diversity, while the increase in organic structure and variety of niches tends to increase it. As we have already pointed out in the discussion of diversity in Chapter 2 there may be an optimum level of diversity for a given energy-flow pattern. We can state that, in general, rapid growth seral stages will tend to have a low diversity on the order of 0.1 or 0.2 on the scale used in Chapter 2, while mature stages will tend to have a higher level on the order of 0.7 or 0.8, unless there is a large energy subsidy that counteracts this pattern.

A decrease in net community production and a corresponding increase in community respiration are two of the most striking and important trends in succession. These changes in community metabolisms are shown graphically in Figure 6-1 which compares ecosystem development in a small laboratory microcosm and in a large natural forest. Total production (P_G) increases faster than energy expenditure (R) at first, so a large net production (P_N) results in a rapid increase in biomass (B). Gradually, equilibrium is established, in about 100 days in the microcosm and 100 or more years in the forest. Perhaps the best way to picture this overall trend is as follows: *Species, biomass, and*

[b] In terms of "information theory" the total information in the community increases as the number of possible interactions between species, individuals, and materials increases.
This table is adapted from "The Strategy of Ecosystem Development," Odum, 1969.

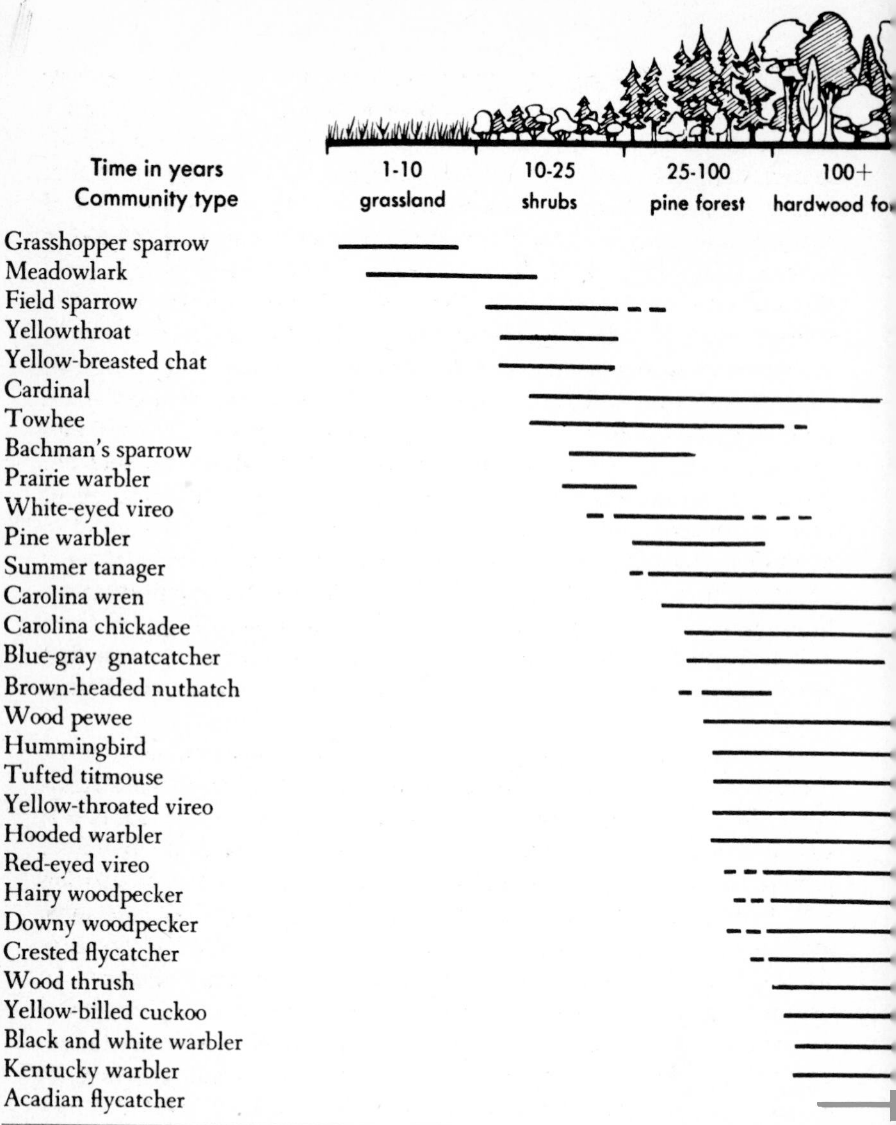

Time in years	1-10	10-25	25-100	100+
Community type	grassland	shrubs	pine forest	hardwood fo

Grasshopper sparrow
Meadowlark
Field sparrow
Yellowthroat
Yellow-breasted chat
Cardinal
Towhee
Bachman's sparrow
Prairie warbler
White-eyed vireo
Pine warbler
Summer tanager
Carolina wren
Carolina chickadee
Blue-gray gnatcatcher
Brown-headed nuthatch
Wood pewee
Hummingbird
Tufted titmouse
Yellow-throated vireo
Hooded warbler
Red-eyed vireo
Hairy woodpecker
Downy woodpecker
Crested flycatcher
Wood thrush
Yellow-billed cuckoo
Black and white warbler
Kentucky warbler
Acadian flycatcher

Number of common species[a]	2	8	15	19
Density (pairs per 100 acres)	27	123	113	233

[a] A common species is arbitrarily designated as one with a density of 5 pairs per 100 acres greater in one or more of the 4 community types.

Fig. 6-2 The general pattern of secondary succession on abandoned farmland in the southeastern United States. The upper diagram shows four stages in the life form of the vegetation (grassland, shrubs, pines, hardwoods) while the bar graph shows changes in passerine bird population that accompany the changes in autotrophs. A similar pattern will be found in any area where a forest is climax, but the species of plants and animals that take part in the development series will vary according to the climate or topography of the area. (Redrawn from D. W. Johnston and E. P. Odum, *Ecology*, Vol. 37, 1956.)

the P/R ratio continue to change long after the maximum gross primary production possible for the site has been achieved. As one evidence for this we may cite the situation in regard to leaves in a terrestrial broad-leaved succession. Agricultural scientists have repeatedly found that maximum productivity of broad-leaved crops occurs when the leaf surface area exposed to the incoming light from above is about 4 or 5 times the surface area of the ground. Any increase in leaves beyond this level does not increase the photosynthetic rate per square meter, since increased shading cancels any advantage that might accrue from increased photosynthetic tissue. In fact, the increased respiration of the extra leaves that do not receive adequate light may reduce the net production of the crop. In a forest the leaf area apparently continues to increase far beyond that limit experimentally shown to increase gross production, since leaf area per ground surface is often 10 or more in an old forest. Since forests are among the most successful of ecosystems with a long geological history of survival, we may well consider the possibility that the extra leaves have other important functions in the ecosystem in addition to production of food. They undoubtedly help moderate temperature and moisture and provide reserves that are important during periods of climatic stress or insect or disease attack.

Natural selection pressure on species in the community shifts dramatically as ecological development proceeds toward the steady-state. These trends are summarized in the bottom section of Table 6-1. Recall from Chapter 5 that *r* was used to designate the potential growth rate of a population and *K* the equilibrium level of population size. In the early stages of succession species with rapid growth rates and ability to exploit unused resources are favored. We can say that the early seral community is under strong "*r*-selection" pressure. In contrast, capacity to live in a crowded world of limited resources is favored in the climax. Larger body sizes, which increase storage capacity, more specialized niches, longer and more complex life cycles, and more cooperation between species (mutualism) are attributes that become more important than reproductive capacity as the ecosystem matures. We can say that the community is now under "*K*-selection" pressures. If a single species is to survive all the way from pioneer systems to mature systems, then dramatic changes must occur in its life-style. As we shall see, this presents a tough challenge to man, whenever he faces the necessity of living at a saturation level.

In theory, then, the "strategy" of ecosystem development involves decreasing entropy, increasing the total information content, and increasing the ecosystem's ability to survive perturbations. Now, let us see what are the constraints against such trends in the varied environments of the biosphere.

ALLOGENIC FORCES AND THE TIME FACTOR IN SUCCESSION

While the changes shown in Figures 6-1, 6-2, and Table 6-1 seem independent of geographical location or type of ecosystem, community structure and physical environment strongly affects: (1) the time required, that is, whether the horizontal scale (x axis, Figures 6-1 and 6-2) is measured in weeks, months, or years; and (2) the relative stability of the climax. In open-water systems, as in cultures, the community is able to modify the physical environment to only a small extent. Consequently, succession in such ecosystems, if it occurs at all, is brief, perhaps lasting for only a few weeks. In a typical marine pond or marine bay, for example, a brief succession from diatoms to dinoflagellates occurs each season, or perhaps several times during a season. A climax, if it can be said to occur, has a limited life span. In a forest ecosystem, to take the other extreme, a large biomass gradually accumulates and the community continues to change in a predictable manner over a long period of time, unless the autogenic processes are interrupted by severe storms or earthquakes. Terrestrial communities developing on sites with level topography and stable substrate provide the best examples of long-term autogenic development, as illustrated in Figure 6-2. The large biological structure that develops in such situations is able to buffer the physical environment and to change the substrate and microclimate to a much greater extent than is possible in a marine situation or a small pond where watershed erosion "pushes," as it were, the succession in a different direction than would occur in the absence of external forces. Where climates are severe, as in deserts or tundras, or steep slopes, as in mountains, autogenic development is more limited and the climax less stable.

A seabeach is a good place to observe the interplay of autogenic and allogenic processes. As long as the wave action is gentle and the sand budget balanced—that is, as much sand is deposited on the average as is removed by tides and waves—the winds build up sand dunes and vegetation develops on them in an orderly sequence something as follows: beach grasses to hardy forbs to woody shrubs to trees such as junipers, pines, and oaks. This community development gradually stabilizes the dunes so that they are resistant to ordinary high tides and occasional storms. However, if the sand budget becomes negative, perhaps resulting from a change in off-shore currents or sand bars (or perhaps because of dredging activities of man), or if storm action increases, or if the sea level rises slightly, then the beach may begin to shift landward and the dunes may start to erode despite the vegetative cover. The dunes then become a source of sand to "renourish" and maintain the beach strand. Only recently have scientists begun to understand this interaction of the geophysical and biological forces. In the past building expensive seawalls, groins, or other artificial barriers was thought to be the answer to beach erosion problems. In many cases this has proved not only to be futile, but it may actually hasten the erosion of the beach because: (1) all the wave forces are now directed at the beach itself, and (2) the source of sand from dunes is cut off by the obstructions. A more prudent procedure is to recognize the natural inherent instability of low-lying shores and regulate development of man-made structures accordingly.

It follows from what we have outlined that more mature successional stages will be more resistant, but by no means immune, to periodic surges in physical forces. Thus, a one-year drought has a very great effect on an early stage of succession or it may completely wipe out a crop of corn or wheat, but it will have much less effect on a

Fig. 6-3 A pulse-stabilized ecosystem in which twice-daily tides maintain an approximate steady state with a large net production as characterizes "youthful" communities. (Georgia salt marshes near Sapelo Island; photo by the author.)

climax forest or grassland. Only if the drought continued for several years would the climax begin to show appreciable changes. In the case of grassland, Dr. J. E. Weaver and his associates at the University of Nebraska have described in detail the changes in species structure and density of the stand that occur in a series of dry years; in general, the mature grassland tends to be set back to a somewhat earlier successional stage containing more annuals and short-lived perennials. However, a rapid recovery occurs on the return of a wet cycle. During the severe droughts of the mid-1930s on the Great Plains of the United States, healthy, mature grasslands, although stressed, were able to survive as intact communities and to hold down the soil, in sharp contrast to the complete biological collapse and severe wind erosion that occurred on croplands and overgrazed areas. For more on this see Weaver and Albertson, 1956.

AGING, THE CYCLIC CLIMAX, AND THE PULSE-STABILIZED SUBCLIMAX

Even without external perturbations the climax does not necessarily remain unchanged forever. Observations in very old forests suggest that self-destructive biological changes may be occurring, which, in the individual, we would call aging. Thus, young trees may not be quite replacing the old ones as they die, or regeneration of nutrients may be lagging and the whole metabolism thus slowing down. There is little data at present, but we wonder if communities may not suffer gradual aging after reaching maturity, just as do individual organisms. Storms and disease, of course, could hasten the aging and death of a climax and the start of a new cycle of developmental stages. In fact, a *cyclic climax* may be a common phenomenon. The California chaparral vegetation mentioned in Chapter 4 is a good example. This dwarf woodland almost seems to "program itself" for periodic destruction by fire. As the community matures, litter and dead wood pile up faster than they can be decomposed during the long, dry summers. Antibiotic chemicals produced by the shrubs also accumulate in the soils and inhibit growth of ground cover. As the community becomes more and more combustible, fire sooner or later sweeps through the woodland. Detritus is removed, antibiotics neutralized, and the shrubs and trees killed back down to ground level. A successional development then repeats itself as the woody vegetation resprouts and grows to maturity again. In this way the aging community becomes youthful again for a while.

So far we have emphasized the *de*stabilizing effect of allogenic physical surges. But acute perturbations can also be stabilizing if they come in the form of regular pulses that can be utilized by adapted species as an extra energy subsidy. In fact, a rhythmic, short-term perturbation imposed from without (as a forcing function in model terminology) can maintain an ecosystem in some intermediate point in the developmental sequence, resulting in, so to speak, a compromise between youth and maturity. What we called "fluctuating water level ecosystems" in Chapter 3 are examples. Estuaries, intertidal shores, rice paddies, and Florida Everglades are held in a highly productive early seral stage by daily or seasonal fluctuations in water levels to which the biota are strongly adapted and coupled in terms of life cycles. These *pulse-stabilized subclimaxes* (by "subclimax" we mean a developmental stage below, or short of, the climax that would develop in the absence of the purturbation) are very important components of the general landscape because the surplus net production that is a property of young systems passes into and helps nourish neighboring systems. This is one reason why ecologists are generally united in recommending that estuaries be preserved and utilized in their more or less natural state.

We might even say that the fire perturbation in the chaparral is actually a stabilizing influence over the long term, since a considerable portion of the flora and fauna would become extinct if the cycle of periodic rejuvenation were not maintained. Because cycles such as dune building and erosion, alternate rising and falling of water, or growth and burning of grasslands or woodlands are inconvenient for man he has a hard time understanding that such things as tides and fire are not necessarily forces he should waste previous fuel energy and money trying to confront. A better idea is to design with the forces ["design with nature" as McHarg (1969) would express it], rather than against them. We wonder how much longer society can afford expensive flood and beach control projects that are for the convenience of small, special interest groups, especially since so many projects in the long run increase flood and erosion damage thereby requiring even more expensive remedial repairs.

THE SIGNIFICANCE OF [E]LOGICAL DEVELOPMENT TO ENVIRONMENTAL MANAGEMENT BY MAN

The principles of ecological ecosystem development are of the greatest importance to mankind. Man must have early successional stages as a continuous source of food and other organic products, since he must have a large net primary production to harvest; in the climax community, because production is mostly consumed by respiration (plant and animal), net community production in an annual cycle may be zero. On the other hand, the stability of the climax and its ability to buffer and control physical forces (such as water and temperature) are desirable characteristics from the viewpoint of the human population. The only way man can have both a productive and a stable environment is to ensure that a good mixture of early and mature successional stages are maintained, with interchanges of energy and materials. Excess food produced in young communities helps feed older stages that in return supply regenerated nutrients and help buffer the extreme of weather (storms, floods, and so on).

In the most stable and productive natural situation there is usually such a combination of successional stages. For example, in areas such as the inland sea of Japan or Long Island Sound, the young communities of plankton feed older, more stable communities on the rocks and on the bottom (benthic communities). The large biomass structure and diversity of the benthic communities provide not only habitat and shelter for life-history stages of pelagic forms but also regenerated nutrients necessary for continued productivity of the plankton. A similar, favorable situation exists in many terrestrial landscapes where productive croplands on the plains are intermingled with diverse forests and orchards on the hills and mountains. The crop fields are, ecologically speaking, "young nature" in that they are maintained as such by the constant labor of the farmer and his machines. The forests represent older, more diverse, and self-sustaining communities that have lower rates of net production but do not require the constant attention of man. It is important that both types of ecosystems be considered together in proper relation. If the forests are destroyed merely for the temporary gain in wood production, water and soil may wash down from the slopes and reduce the productivity of the plains. Ruins of civilizations and man-made deserts in various parts of the world stand as evidence that man has not been fully aware of his need for protective as well as productive environments. Mature systems have other values to mankind in addition to products; they should not be considered as crops in the sense of wheat or corn. The conservationist speaks of a policy of balancing contradictory needs as "multiple use," but in the past he has found it difficult to translate long-term values into monetary units. Consequently, too often the possibilities for immediate economic gain in harvest overrides what later turns out to be a more important value.

To illustrate these difficulties, let us consider the controversy over National Forest management policy that received considerable public attention in the early 1970s. For the most part National Forests have been managed on a "selective-cut" basis; that is, selected trees, including a portion of mature trees that are no longer growing, are removed periodically leaving the stand more or less intact to serve other uses (recreation, soil and water stabilization, and so on) and leaving room for younger trees to grow faster. As the demand for paper and other wood products became acute there was pressure for harvesting on a "clear-cut and replant" cycle, since the yield would then be greater and subsequent rate of net production increased. But right after the clear-cut the system would be subject to various disorders such as soil erosion and nutrient loss; the cost of taking care of these problems could cancel out the value of extra wood yield. Thus, both plans have advantages and disadvantages. A sensible solution to the dilemma would be to vary the management according to the site capability. Where topography is steep and soil thin, or where the vegetation is botanically unique, or of great scenic beauty, a selective-cut plan would be best in the long run. Where topography is more level, the soil deep and stable, and the species capable of rapid regrowth, then a clear-cut procedure could be a desirable choice.

In essence there are only two basic ways to meet the problem of youth and maturity in the landscape. One would be to maintain intermediate states as naturally occurs in pulse-stabilized systems, and the other would be to compartmentalize or "zone" the landscape so as to have separate areas primarily managed for production and for protection. Both require that society adopts regional land-use plans, an idea whose time is coming. We will come back to this in Chapter 8. For now, we simply conclude that the principles of ecosystem development provide important natural guidelines for determining options and making decisions as to how to make optimum use of the total environment.

EVOLUTION OF THE ECOSYSTEM

As in the case with short-term development, as described earlier in this chapter, the long-term evolution of ecosystems is shaped by the interaction of allogenic geological and climatic changes and autogenic processes resulting from the activities of the living components of the ecosystem. In a broad sense the "strategy" of long-term evolutionary development is the same as that of short-term ecological succession, namely, increased control of, or homeostasis with, the physical environment in the sense of achieving maximum protection from its perturbations.

Although we may never know exactly how life began on earth, the generally accepted theory is that the first living things were tiny anaerobic (living without free oxygen) heterotrophs that lived on organic matter synthesized by abiotic processes. The first successional development, then, may have been more like the hay infusion culture model (see page 154) than the autotrophic culture model. The atmosphere at the time of the origin of life 3 billion years ago contained nitrogen, hydrogen, carbon dioxide, water vapor, but little or no oxygen. It also contained carbon monoxide, chlorine, and hydrogen sulfide in quantities that would be poisonous to much of present day life. The composition of the atmosphere in those early days was largely determined by the gaseous stuff that comes out of volcanos. The geologist would speak of this as "atmospheric formation by crustal outgassing." The earth's early reducing atmosphere (a term to contrast with oxygenic atmosphere) may have been similar to that now found on Venus or Jupiter. Because of the lack of gaseous oxygen there was no ozone layer, as there is now. Molecular oxygen O_2, acted on by short-wave ultraviolet radiation produces ozone, or O_3, which in turn shields out the deadly radiation. Thus, at first, life could exist only if shielded by water or other barriers, but strange to say it was the short-waved radiation that is thought to have created a chemical evolution leading to complex organic molecules such as amino acids that became the building blocks of life. This synthesis also provided food for the first organisms.

For millions of years life apparently remained as only a tiny foothold, limited in habitat and energy source, in a violent physical world. The big change began with the appearance of the first photosynthetic algae which were able to make food from simple inorganic substances and which released gaseous oxygen as a by-product. As the oxygen diffused into the atmosphere, the ozone shield developed and life could then spread to all parts of the globe, and there followed an almost explosive evolution of increasingly complex aerobic organisms. The broad pattern of the evolution of organisms and the oxygenic atmosphere that make the biosphere absolutely unique in our solar system is shown in Figure 6-4. Over long stretches of time production exceeded respiration ($P/R > 1$) so oxygen increased and CO_2 decreased. Our fossil fuels were also formed during periods when P exceeded R by a wide margin.

Incidentally, I can think of no better way to dramatize man's dependence on his environment and his need to become a wise custodian of this frail earth than to recount how our atmosphere came into being, emphasizing, of course, that it was built by microorganisms, not by men. I think the story of our air should be told to every schoolchild and every citizen. It is a fascinating drama of living history with enough mystery and potential tragedy to intrigue teacher and pupil alike. It is a subject that lends itself to student participation in learning since the possibilities for study projects, artwork, plays, and the

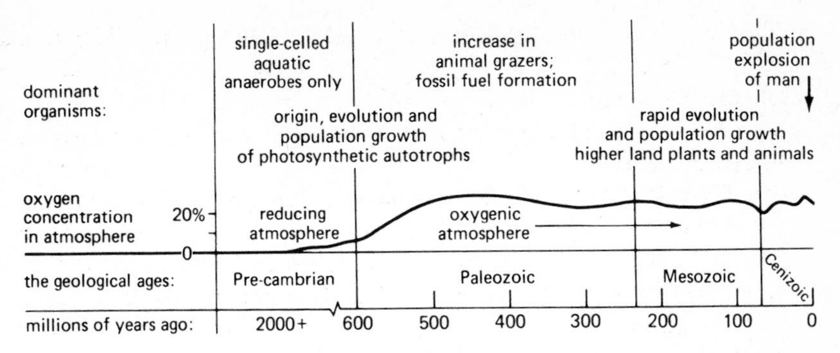

Fig. 6-4 The evolution of the biosphere in terms of the oxygenation of the atmosphere as linked with the evoluation of the biota.

like are unlimited. Berkner and Marshall have written both a popular account (1966) and a more technical treatise (1964) that provide good reference.

As was noted in Chapter 5, the origin of new species and the evolution of larger and more complex multicellular organisms is believed to occur principally through natural selection at the population level. However, there is increasing evidence that evolution at the community and ecosystem level is also important. One way this can occur is by reciprocal selection between interdependent species by the process known as *coevolution*. The best documented cases involve the coupling of autotrophs and heterotrophs. In Mexico, Janzen (1966) describes how an ant and a species of Acacia tree have evolved together to such an extent that neither can survive without the other. The lichen and the other examples of "mutualism" noted in Chapter 5 provide other possible examples. As one species evolves to take a selective advantage of the association, the other species comes under selective pressure to strengthen the interdependence. We can see how a *genetic feedback* process of this sort could shape the evolution of a whole ecosystem.

Another possibility is what is known as *group selection*, or natural selection between groups of organisms that are not closely linked by mutualistic associations. Group selection leads theoretically to the maintenance of traits favorable to the community as a whole even though selectively disadvantageous to the genetic carriers within the populations; the common good wins out over the individual good, so to speak. Conversely, group selection may eliminate traits unfavorable at the group level even though favorable at the species level. Group selection, while appealing to the ecologist, is a highly controversial subject among geneticists, so we had best leave it for more study.

SUGGESTED READINGS

References cited

Berkner and Marshall. See citations, Chapter 4.

Beyers, Robert J. 1964. The microcosm approach to ecosystem biology *Amer. Biol. Teacher.* 26:491–498.

Cooke, G. Dennis. 1967. The pattern of autotrophic succession in laborator microecosystems. *Bio-Sci.* 17:717–721.

Janzen, D. H. 1966. Coevolution of mutualism between ants and acacias i Central America. *Evol.* 20:249–275.

Johnston, D. W. and E. P. Odum. 1956. Breeding bird populations in rela tion to plant succession on the Piedmont of Georgia. *Ecol.* 37:50–62

Kira, T. and T. Shidei. 1967. Primary production and turnover of organi matter in different forest ecosystems of the western Pacific. *Jap. Jour Ecol.* 17:70–87.

McHarg, Ian L. 1969. *Design with Nature.* Garden City, New York: Natura History Press.

Odum, Eugene P. 1969. The strategy of ecosystem development. *Science* 164:262–270.

Weaver, J. E. and Albertson, F. W. Grasslands of the Great Plains; Thei Nature and Use. Johnsen Publ. Co. Lincoln, Nebraska.

Ecosystem development (ecological succession)

Clements, F. E. and V. E. Shelford. 1939. *Bioecology.* New York: John Wiley & Sons. (Parallel succession of plants and animals is a majo theme of this book.)

Dolan, Robert; Paul J. Godfrey; and William E. Odum. 1973. Man's impact on the Barrier Islands of North Carolina. *Amer. Sci.* 61(2):152–162. (Interesting, beautifully illustrated account of autogenic and allogenic impacts on beaches.)

Drury, W. H. and I. C. T. Nisbet. 1973. Succession. *Jour. Arnold Arboretum.* 54: 331–368 (Authors review population-level and community-level theories and opt for the former.)

Margalef, Ramon. 1968. *Perspectives in Ecological Theory.* Chapter 2. Chicago: Univ. of Chicago Press. (See also his longer paper on succession in: *Advanced Frontiers of Plant Science.* Inst. Adv. Sci. and Culture, New Delhi, India. 2:137–374.)

Odum, Eugene P. 1969. The Strategy of Ecosystem Development. *Science.* 164:262–270.

———. 1971. *Fundamentals of Ecology,* 3rd ed. Chapter 9. Philadelphia: Saunders.

Olson, J. S. 1958. Rates of succession and soil changes on southern Lake Michigan sand dunes. *Bot. Gazette.* 119:125–170. (Reexamination of succession on sand dunes as first described by Cowles in 1899.)

Whittaker, Robert H. 1970. Communities and ecosystems. New York: Macmillan.

Evolution of the ecosystem

Cloud, Preston E. 1974. Evolution of ecosystems. *Amer. Sci.* 62:54–66.

Dobzhansky, Th. 1950. Evolution in the tropics. *Amer. Sci.* 38:209–221.

Ehrlich, Paul R. and Peter H. Raven. 1964. Butterflies and plants: a study in coevolution. *Evol.* 18:586–608.

Fischer, Alfred G. 1972. Atmosphere and the evolution of life. *Main Currents Modern Thought.* 28(5), May-June.

Margulis, Lynn. 1971. Symbiosis and evolution. *Sci. Amer.* 225(2):48–57.

Solbrig, Otto T. 1971. The population biology of dandelions. *Sci. Amer.* 59:686–694. (*r*-strategist and *k*-strategist genetic varieties adapt this plant to disturbed and undisturbed conditions, respectively.)

Whittaker, R. H. and G. M. Woodwell. 1972. Evolution of natural communities. In *Ecosystem Structure and Function,* ed. J. A. Weins, pp. 137–156. Corvalis: Oregon State Univ. Press.

Wilson, Edward O. 1973. Group selection and its significance for ecology. *Bio-Sci.* 23:631–638.

chapter 7

Major Ecosystems of the World

For the most part in this book we have based our approach to ecology on the analysis of units of the landscape as ecological systems. Principles and common denominators that apply to any and all situations, whether aquatic or terrestrial, natural or man-made, have been emphasized. The importance of the driving force of energy has been stressed. In Chapter 5 another useful approach was introduced, that of concentrating study on population units which are the vehicles for evolutionary change. Still another useful approach is geographical—involving the study of the pattern of earth forms, climates, and biotic communities that make up the biosphere. In this chapter we shall list and briefly characterize the major ecological formations, or easily recognized ecosystem types, with emphasis on geographical and biological differences that underlie the remarkable diversity of life on earth. In this manner we hope to establish a *global* frame of reference for the next and final chapter, which deals with mankind's new challenge to attack his problems on a large scale.

We would do well to start our world tour with the seas, the largest and most stable ecosystem. The sea, presumably, was the first ecosystem, for life is now thought to have originated in the saltwater milieu.

THE SEAS The major oceans (Atlantic, Pacific, Indian, Arctic, and Antarctic) and their connectors and extensions cover approximately 70 percent of the earth's surface. Physical factors dominate life in the ocean (Figure 7-1A). Waves, tides, currents, salinities, temperatures, pressures, and light intensities largely determine the makeup of biological communities that, in turn, have considerable influence on the composition of bottom sediments and gases in solution. The food chains of the sea begin with the smallest known autotrophs and end with the largest of animals (giant fish, squid, and whales). The study of the physics, chemistry, geology, and biology of the sea are combined into a sort of "superscience" called *oceanography,* which is becoming increasingly important as an international force. Although exploration of the sea is not quite as expensive as exploration of outer space, a considerable outlay of ships, shore laboratories, equipment, and specialists are needed. Most research is of necessity carried out by a relatively few large institutions backed by government subsidies, mostly from the affluent nations.

To fully appreciate both the promise and the problems involved in man's use of the sea we need to look at the contour of the sea bottom, as shown in Figure 7-1C, which also gives standard oceanographic nomenclature for zones of the sea. According to the now widely accepted "continental drift theory," some of the continents, especially Africa and South America as one pair and Europe and North America as another, were once quite close together and have drifted apart through the ages. The mid-Atlantic ridge (Figure 7-1C) is, according to this theory, the line of former contact between continents now hundreds of miles apart. As a citizen you will be hearing a lot about the continental shelf, that sloping plateau that borders the continents. Located here are the bulk of undersea oil and mineral wealth. From the edge of the shelf, which varies greatly in width from location to location, the continental slope drops off rapidly into the true depth of the sea. The topography of the continental slope is very rugged with huge

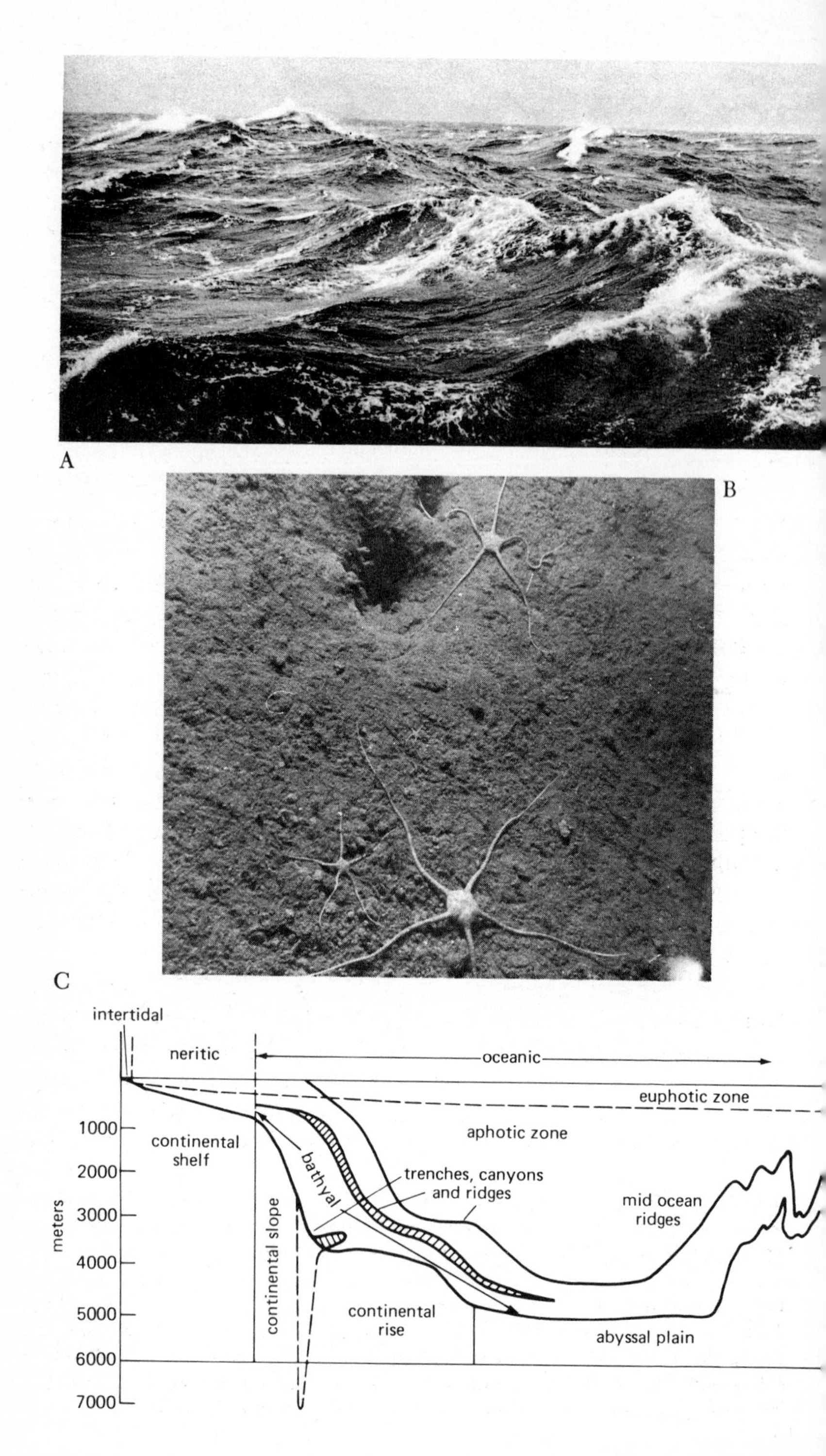
A
B
C
intertidal
neritic
oceanic
euphotic zone
aphotic zone
1000
2000
3000
4000
5000
6000
7000
meters
continental shelf
bathyal
trenches, canyons and ridges
mid ocean ridges
continental slope
continental rise
abyssal plain

canyons and ridges that are constantly changing under the forces of volcanic action and underwater "landslides."

Since there are likely to be phytoplankton under every square meter and since life in some form extends to the greatest depths, the seas are the largest and "thickest" of ecosystems. They are also biologically the most diverse. Marine organisms exhibit an incredible array of adaptations, ranging from flotation devices that keep the tiny plankters within the upper layers of water, to the huge mouths and stomachs of deep-sea fish that live in a dark, cold world where meals are bulky but few and far between. As shown in Figure 3-7, the continental shelf areas are fairly productive; seafood harvested here is an important source of protein and minerals for man. The most productive areas and largest fisheries are those that benefit from nutrients carried up by upwelling currents, a form of energy subsidy. Strong upwelling occurs in certain areas along the west coasts of the several continents. The Peruvian upwelling region, one of the most productive natural areas in the world, was singled out for special discussion on page 80. The vast stretches of the deep sea, however, are mostly semidesert with considerable total energy flow (because of the large area) but not much per unit of area. The autotrophic layer (photic zone) is so small in comparison with the heterotrophic layer (see Figure 7-B) that the nutrient supply in the former is limiting (see Chapter 3). A number of schemes have been proposed, and there are now several experiments under way, to tap the potential energy of vertical temperature differences to create artificial upwelling. Even if man is not able to obtain much food from the deep-sea area it is nevertheless very important to him, for the seas act as a giant regulator that helps to moderate land climates and maintain favorable concentrations of carbon dioxide and oxygen in the atmosphere.

International conferences are now being scheduled to discuss the thorny problem of setting up international law with rules and regulations for exploiting seabed minerals and energy resources. Since, as we have noted, most mineral wealth, as well as most exploitable food, is located near shore, it would seem reasonable for each country to assume stewardship of the shelf area adjacent to its land territory, but

Fig. 7-1(A) The seas. The never-ending wave motion seen in the photograph serves to emphasize the dominance of physical factors in the open ocean. (Courtesy Woods Hole Oceanographic Institute and D. M. Owen.)
(B) In many places the bottom of the sea (in contrast to surface) is a relatively quiet and stable environment. The photograph shows an area about 17 by 20 in. at a depth of 1500 m. on a transect between Cape Cod and Bermuda in the Atlantic. Several brittle starfish are visible as well as worm tubes and two large worm burrows. (Photo courtesy Woods Hole Oceanographic Institution). (C) Zonation and bottom contour of the Atlantic.

the large differences in width of the shelf make such a simple solutio difficult, perhaps impractical. Most objective assessments (see, fo example, Cloud, 1969) warn against undue optimism that the deep se is a vast storehouse just waiting to be exploited. Recovering suc resources as there are will be even more expensive than getting mir erals and oil from the shelf where costs are indeed huge. Remembe that the sea is more important as a life support and climate regulato than it is as a supply depot. Anything we do to exploit the latter mus not jeopardize the former (recall our point about "gross and net energy, page 84).

ESTUARIES AND SEASHORES Between the seas and the continents lies band of diverse ecosystems that are not jus transition zones but have ecological characteristics of their own Whereas physical factors such as salinity and temperature are muc more variable near shore than in the sea itself, food conditions ar so much better that the region is packed with life. Along the shor live thousands of adapted species that are not to be found in the ope sea, on land, or in fresh water. A rocky shore, a sand beach, an inter tidal mud flat, and a tidal estuary dominated by salt marshes are show in Figure 7-2 to illustrate four kinds of marine inshore ecosystems. Th word "estuary" (from Latin *aestus,* tide) refers to a semienclosed bod of water, such as a river mouth or coastal bay where the salinity i intermediate between the sea and fresh water, and where tidal actio is an important physical regulator and energy subsidy (see footnot to Table 2-1).

Estuaries and inshore marine waters are among the most natu rally fertile in the world. Three major life forms of autotrophs are often intermixed in an estuary and play varying roles in maintaining a high gross production rate; these are: (1) phytoplankton; (2) ben thic microflora—algae living in and on mud, sand, rocks or other hard surfaces, and bodies or shells of animals; and (3) macroflora—large attached plants—the seaweeds, submerged eelgrasses, emergent marsh grasses, and, in the tropics, mangrove trees. An estuary is ofter an efficient nutrient trap that is partly physical (differences in salini ties retard vertical but not horizontal mixing of water masses) and partly biological, as was illustrated by the example of the mussel population (Figure 5-8). As discussed in the next chapter, this property enhances the estuary's capacity to absorb nutrients in wastes provided organic matter has been reduced by secondary treatment. Estuaries provide the "nursery grounds" (that is, place for young stages to grow rapidly) for most coastal shellfish and fish that are harvested not only in the estuary but offshore as well.

Organisms have evolved many adaptations to cope with tidal cycles, thereby enabling them to exploit the many advantages of living in an estuary. Some animals, such as fiddler crabs, have internal "biological clocks" that help to time their feeding activities to the most favorable part of the tidal cycle. If such animals are experimentally removed to a constant environment they continue to exhibit rhythmic activity synchronous with the tides. Estuaries have been traditionally the most used, but least appreciated, free sewers for man's great coastal developments. As symptoms of overuse appear (the decline in seafood yield is often a first symptom) government becomes concerned with "coastal management." An economic approach to proper evaluation of estuaries is discussed in the next chapter.

STREAMS AND RIVERS The history of man has often been shaped by the rivers that provide water, transportation, and a means of waste disposal. Although the total surface area of rivers and streams is small compared to that of oceans and land mass, rivers are among the most intensely used by man of natural ecosystems. As in the case of estuaries, the need for "multiple use" (as

Fig. 7-2 Four types of coastal ecosystems. (A) A rocky shore on the California coast, characterized by underwater seaweed beds, tidepools with colorful invertebrates, sea lions ("seals") and sea birds (seen in the water and on rocks off shore). (B) A sand beach with ghost crab near its burrow. (C) An intertidal mud flat in Massachusetts, during low tide. Although mud flats may look like deserts on the surface they can, when not polluted or overexploited by man, support very large populations of shellfish and other animals. Shown in the picture is Dr. Paul Galtsoff, a lifelong student of marine clams, with some 30-odd clams dug from a single square foot (to depth of 8 in.) of mud flat. Animals of mud flats are of two general feeding types: filter feeders such as the clams, which filter out food particles from water, and deposit feeders such as many gastropods that ingest the "mud" from which organic matter is extracted in the intestine. Part of the food that supports the dense populations is produced by algae living on and in the mud, and part is brought in by each tide. (D) A productive tidal estuary on the coast of Georgia, showing sounds, networks of tidal creeks, and vast areas of salt marsh. The shallow creeks and marshes not only support an abundance of stationary organisms but they also serve as nursery grounds for shrimp and fish that later move off shore where they are harvested by trawlers. [(A), (B), and (C), U. S. Department of Interior, Fish and Wildlife Service: (D), University of Georgia Marine Institute.]

A

C

D

contrasted to a "single use" approach to such ecosystems as cropland demands that the various areas (water supply, waste disposal, fis production, flood control, and so on) be considered together and n as entirely separate problems.

From the energetic standpoint rivers and streams are incon plete ecosystems; that is, some portion, often a large portion, of th biological energy flow is based on organic matter imported fro adjacent terrestrial ecosystems, or sometimes from adjacent lakes (se page 72). Although streams are naturally adapted waste treatme systems for degradable wastes (recall our frequent comment abo "free sewers") almost all of the world's great rivers are severely ove loaded with the residues of man's civilization. As geographer M. C Wolman (1971) has concluded, "demands on water resources a increasing at a rate that exceeds the rate of installation of waste trea ment facilities." This is another one of those "mismatched rates" th are at the heart of man's troubles with his environment. In all par of the world man has so extensively dammed, diked, and channelize rivers that it is getting hard to find a truly wild river of any size. is turning out that some of these manipulations bring only tempora or local benefits at great cost, and create additional problems costin still more money to correct (as in the case of some flood contr projects). Accordingly, flood damages that used to be considere "natural disasters" (and, therefore, unavoidable) are more and mo proving to be man-made disasters (and, therefore, avoidable). I the future, proposed alterations will have to be subjected to a mo thorough cost-benefit analysis than was the case in the past. Mo about this in the next chapter.

The stream ecologist finds it convenient to consider flowin water ecosystems under two subdivisions: (1) streams in which th basin is eroding and the bottom, therefore, is generally firm; and (2 streams in which material is being deposited and, therefore, the bo tom is generally composed of soft sediments. In many cases thes situations alternate in the same stream, as may be seen in the "rapid and "pools" of small streams. Aquatic communities are quite differe in the two situations owing to the rather different conditions of exis ence. The communities of pools resemble those of ponds in that considerable development of phytoplankton may occur and the speci of fish and aquatic insects are the same or similar to those found i ponds and lakes. The life of the hard-bottom rapids, however, is con posed of more unique and specialized forms, such as the net-spinnin caddis (larvae of insects called caddis flies or Trichoptera), whic constructs a fine silk net that removes food particles from the flowin waters.

The load of sediments discharged into the oceans by the great rivers of the world tell us something about man's treatment of the land. The rivers of Asia, the continent with the oldest civilizations and the most intense human pressure on the land, discharge 1500 tons of soil per square mile of land area annually. In contrast, the sediment discharge rate for North America is 245, South America 160, and Europe 90 (data from Holeman, 1968).

LAKES AND PONDS In the geological sense, most basins that now contain standing fresh water are relatively young. The life span of ponds ranges from a few weeks or months in the case of small seasonal ponds to several hundred years for larger ponds. Although a few lakes, such as Lake Baikal in Russia, are ancient, most large lakes date only as far back as the ice ages. Standing-water ecosystems may be expected to change with time at rates more or less inversely proportional to size and depth. Although geographical discontinuity of fresh waters favors speciation, the lack of isolation in time does not. Generally speaking, the species diversity is low in freshwater communities and many taxa (species, genera, families) are widely distributed within a continental mass and even between adjacent continents. A pond was considered in some detail in Chapter 2 as an example of a convenient-sized ecosystem for introducing the study of ecology.

Distinct zonation and stratification are characteristic features of lakes and large ponds. Typically, we may distinguish a *littoral zone* containing rooted vegetation along shore, a *limnetic zone* of open water dominated by plankton, and a deep-water *profundal zone* containing only heterotrophs. These zones parallel the major zone of the sea, as shown in Figure 7-1B. In temperate regions, lakes often become thermally stratified during summer and again in winter, owing to differential heating and cooling. The warmer upper part of the lake, or *epilimnion* (from Greek *limnion*, lake) becomes temporarily isolated from the colder lower water, or *hypolimnion*, by a *thermocline* zone that acts as a barrier to exchange of materials. Consequently, the supply of oxygen in the hypolimnion and nutrients in the epilimnion may run short. During spring and fall, as the entire body of water approaches the same temperature, mixing again occurs. "Blooms" of phytoplankton often follow these seasonal rejuvenations of the ecosystem. In warm climates mixing may occur only once a year (winter), while in the tropics mixing is a gradual or irregular process.

Primary production in standing-water ecosystems depends o the chemical nature of the basin, the nature of imports from stream or land, and the depth. Shallow lakes are usually more fertile tha deep ones for reasons already outlined in the discussion of the sea (see also Chapter 3 and Figure 3-7). Accordingly, the yield of fish pe acre of surface is generally inversely proportional to the mean deptl Lakes are often classified into *oligotrophic* ("few foods") and *eutro phic* ("good foods") types depending on their productivity. What ha now come to be known as "*artificial or cultural eutrophication*" c lakes has created difficult problems in the vicinity of metropolita areas and crowded resorts. Inorganic fertilizers in sewage effluer entering lakes increases their primary production rates and change the composition of the aquatic community in ways that are not popul with the public. For example, game fish such as trout, which requir cool, clear, oxygen-rich waters, may disappear; growth of algae an other aquatic plants may become so great as to interfere with swim ming, boating, and sport fishing; or undecomposed dissolved organic may impart a bad taste to water even after it has passed throug water purification systems. Thus, a biologically poor lake is preferabl to a fertile one from the standpoint of water use and recreation. Agai we have a paradox. In some parts of the biosphere man is doin everything possible to increase its fertility in order to feed himsel whereas in other places he does everything possible to prevent fe tility (by removing nutrients, poisoning plants, and so on) in orde to maintain a pleasant environment. A fertile green pond capable c producing many fish is not considered to be a good recreation swimming pool.

In recent years, efforts to divert municipal wastes from certai lakes has demonstrated that cultural eutrophication can be reverse in the sense that some lakes will return to a less fertile condition wit improved water quality (in terms of human use) when nutrients n longer pour into them. Lake Washington in Seattle is a wel documented case (see Edmondson, 1968).

Constructing artificial ponds and lakes (impoundments) is on of the conspicuous ways in which man has changed the landscape i regions that lack natural bodies of water. In the United States almo every farm now has at least one farm pond, and large impoundmen have been constructed on practically every river. Most of this activit works to the benefit of both man and the landscape, for water an nutrient cycles are stabilized and the added diversity is a welcom change in man's usual tendency to create a monotonous landscap However, the impoundment idea can be carried too far; covering u fertile land with a body of water that cannot yield much food ma not be the best land use. The need for broader analysis of presumed benefits of impoundments was mentioned in the preceding section.

People seem strangely unprepared for the changes that arise from ecological succession in artificial ponds and lakes. Somehow, once a lake has been created it is expected to remain the same as would a skyscraper or bridge. Instead, of course, all the processes of succession that were described in Chapter 6 take place with changes resulting from activities of the biotic community (autogenic processes) and, especially in ponds and shallow lakes, changes resulting from sediment discharges from the watershed (allogenic processes).

Fishing is often very good for the first few years in a new impoundment, then declines as the excess nutrients in the flooded watershed are exploited and the body of water begins to age. Figure 7-3 graphs estimated fish abundance based on two catch methods from the second to the fifteenth year after impoundment of water in a large,

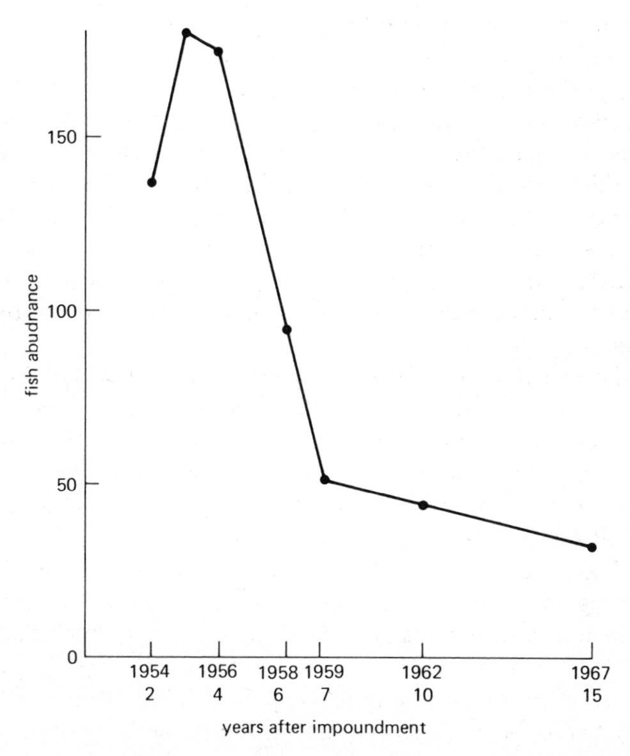

Fig. 7-3 Fish abundance (based on mean of two sampling methods) in a new mainstream reservoir on the upper Missouri River from second to fifteenth year after completion of the dam and full impoundment of water in lake Francis Case, South Dakota. (Data from Gasaway, 1970.)

A

C

mainstream reservoir on the upper Missouri River. Fish abundance reached a peak in the third and fourth year, but declined to less than one-third the level thereafter. Ponds and small lakes can often be "rejuvenated" by drawing down the water or periodically draining and refilling so that body of water is maintained in a young or early successional stage. Such "fallowing" of ponds has been practiced for centuries in Europe and in the Orient. Fluctuation of water level provides an energy subsidy, as outlined on page 163.

Since shallow bodies of water can be as productive as an equal area of land, *aquaculture* can be a useful supplement to agriculture, especially for production of high protein food. Aquaculture is a highly developed art and science in oriental countries where large yields of algae, fish, and shellfish are obtained from managed but seminatural bodies of both fresh and salt water. The approach to fish culture, interestingly enough, is greatly influenced by population density. Where man is crowded and hungry, ponds are managed for their yields of herbivores such as carp; yields of 1000 to 5000 lb/acre per year can be obtained (more if supplemental food is added). Where man is not crowded or hungry, sport fish are what is desired; since these fish are usually carnivores produced at the end of long food chains, the yields are much less, 100 to 500 lb/acre/year.

FRESH-WATER MARSHES Much of what was said about estuaries also applied to freshwater marshes (Figure 7-4A); they tend to be naturally fertile ecosystems. Tidal action, of course, is absent, but periodic fluctuation in water levels resulting from seasonal and annual rainfall variations often accomplishes the same thing in terms of maintaining long-range stability and fertility. Fires during dry periods consume accumulated organic matter thereby deepening the water-holding basins and aiding subsequent aerobic decomposition and release of soluble nutrients, thus increasing the rate of production. In fact, if such events as drawdown and fire do not occur, the build-up of sediments and peat (undecayed organic matter),

Fig. 7-4 Three fresh-water ecosystems. (A) A fresh-water marsh in the Sacramento National Wildlife Refuge in California, where flocks of geese find refuge and shelter in productive aquatic and semiaquatic vegetation. (B) A natural pond in the grassland region of Western Canada. (C) Convergence of two streams in northern New Jersey. The stream in the foreground flows from a watershed protected by grass and trees; the stream entering from the left is badly polluted with silt as a result of poor agriculture. [(A) and (B) U. S. Department of Interior, Fish and Wildlife Service. (C) U. S. Soil Conservation Service.]

tends to lead to the invasion of terrestrial woody vegetation. Where man controls water levels by dikes in marshes he generally finds that chemical herbicides or mechanical methods have to be used if the area is to continue to exist as a true freshwater marsh ecosystem suitable for ducks and other semiaquatic organisms.

The general public prejudice against marshes is understandable, since they are sometimes the home of mosquitoes and other disease carriers and pests. Before much was known about the life history and ecology of the arthropods and snails as disease carriers, destroying their habitat (that is, draining the marsh) was about the only solution. Our present knowledge now makes it unnecessary to destroy the ecosystem in order to control undesirable species.

In addition to producing ducks and fur-bearers, marshes are valuable in maintaining water tables in adjacent ecosystems. The Florida Everglades are an exceptionally large and interesting stretch of freshwater marshes characterized by naturally fluctuating water levels. Complete drainage (even if possible or otherwise desirable) would not only ruin the area as a wildlife paradise but would also be risky in that salt water might then intrude into the underground water supply needed by the large coastal cities. Likewise, complete stabilization of water levels would also destroy the unique features of the Everglades, for reasons given at the beginning of this section.

Finally, it is significant that rice culture, one of the most productive and dependable of agricultural systems yet devised by man, is actually a type of freshwater marsh ecosystem. The flooding, draining, and careful rebuilding of the rice paddy each year has much to do with the maintenance of continuous fertility and high production of the rice plant, which, itself, is a kind of cultivated marsh grass.

THE TERRESTRIAL ORMATIONS, THE BIOMES

Large, easily recognized terrestrial community units are known as *biomes*. In a given biome the life form of the climax vegetation (see Chapter 6 for an explanation of the concept of climax) is uniform, and is the key to recognition. Thus, the dominant climax vegetation in the grassland biome is grass, although the species of dominant grasses will vary in different geographical regions where the grassland biome occurs (see page 46). Other types of vegetation will be included in the biome, as for example, "weedy" seral stages in succession, forest subclimaxes related to local soil and water conditions, crops, and other vegetation introduced by man.

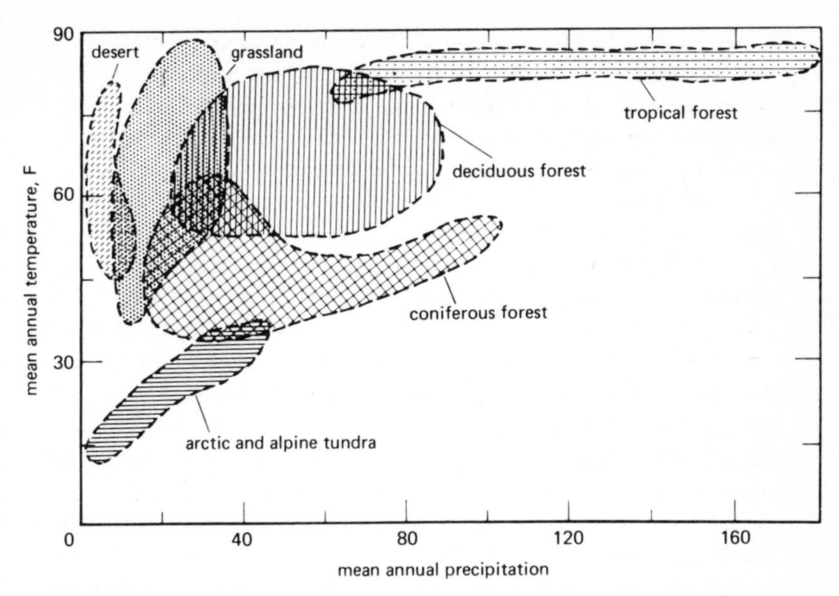

Fig. 7-5 Distribution of six major biomes in terms of mean annual temperature and mean annual rainfall. (Courtesy National Science Foundation.)

The distribution of six major biomes in relation to temperature and rainfall is shown in Figure 7-5. If you will check the mean annual temperature and rainfall of your locality you can determine from Figure 7-5 which biomes you live in, even if you are now sitting in the middle of a city with no climax vegetation anywhere around. Several other biomes (not shown in Figure 7-5), such as chaparral, tropical savanna, thorn shrub, and tropical monsoon forests are related to seasonal distribution of rainfall rather than annual means.

For the past 6 years most nations of the world have taken part in what is known as the "International Biological Program" involving governmental grant support for interdisciplinary team research and systems modelling of major biomes. Both the "modelling up" and the "modelling down" approaches, as mentioned on page 8-9, are being tested in efforts to develop realistic working models that will help man better understand his impact on the natural biome matrix in which his civilization is embedded. For a brief review of the United States program and some of its accomplishments, see Hammond (1974). Other nations have, in general, taken a less holistic approach with varying emphasis on such aspects as natural area inventory, impact of grazing animals, secondary production, detailed descriptive analysis of vegetation, environmental health, and preservation of genetic stocks of endangered wild and domestic organisms.

DESERTS Desert biomes occur in regions with less than 10 in. of annual rainfall, or sometimes in hot regions where there is more rainfall, but unevenly distributed in the annual cycle. Lack of rain in the mid-latitudes is often due to stable high-pressure zones; deserts in temperate regions often lie in "rain shadows," that is, where high mountains block off moisture from the seas. Two types of North American deserts are shown in Figure 7-6, a "cool" desert in Washington, with sage brush, and a "hot" desert in Arizona, where creosote bushes and cacti are conspicuous. The characteristic spacing of desert vegetation and the possibility of "birth control" mechanisms were discussed in Chapter 5. North American deserts are not as extreme as those in other continents, such as the African Sahara or the Asian Gobi. Some seasonal rain can be expected every year in U. S. deserts, but rainless periods in extreme deserts may span years.

Four very distinctive life forms of plants are adapted to the desert ecosystem: (1) The annuals (such as cheat grass, shown in Figure 7-6B), which avoid drought by growing only when there is adequate moisture (see Chapter 4 , page 115). (2) The desert shrub with numerous branches arising from a short basal trunk, and small, thick leaves that may be shed during dry periods; the desert shrub survives by its ability to become dormant before wilting occurs. In the cooler deserts, the shrubs develop very deep root systems that tap moisture that remains available after the surface completely dries out. In such cases the leaves and stems may remain green and active throughout the summer. (3) The succulents, such as the cacti of the New World or the euphorbias of the Old World, which store water in their tissues. (4) Microflora, such as mosses, lichens, and blue-green algae that remain dormant in the soil but are able to respond quickly to cool or wet periods.

Animals such as reptiles and some insects are "preadapted" to deserts, for their impervious integuments and dry excretions enable them to get along on the small amount of water. Mammals as a group are poorly adapted to deserts but some few species have become secondarily adapted. A few species of nocturnal rodents, for example, that excrete very concentrated urine and do not use water for temperature regulation, can live in the desert without drinking water. Other animals such as camels must drink periodically but are physiologically adapted to withstand tissue dehydration for periods of time. For more on adaptations of desert animals, see Schmidt-Nielson (1973).

In the past mankind has developed remarkable cultures, including adapted domestic plants and animals for life in or along the edges of deserts. In fact, life in dry regions requires ingenuity and a conservation ethic, two attributes badly needed in more benign regions.

A

B

Fig. 7-6 Two types of deserts in western North America. (A) A "hot" desert in Arizona. (B) A "cool" desert in eastern Washington in early spring. The desert shrub life form is illustrated by the dark creosote bushes in (A) and the sagebrush in (B). Note the rather even spacing of shrubs, especially evident in the upper picture. The succulent life form is represented by cacti in (A) and the desert annual is represented by the cheat grass growing between the sage bushes in (B). The objective of the radioactive tracer experiment shown in (B) was to determine the relative uptake from soil of specific minerals by the two life forms growing within the metal ring. [(A) Courtesy Dr. R. R. Humphries. (B) Hanford Atomic Products Operation.]

Because water is the dominant limiting factor, the productivity of a given desert region is almost a linear function of rainfall. In the California Mohave desert a 100-mm annual rainfall will result in about 600 kg dry matter/ha while 200-mm will increase net production to about 1000 kg/ha. Where evaporative losses are less in the cooler Great Basin deserts, a 200-mm rain produces 1500–2000 kg/ha.

Where soils are suitable, irrigation can convert deserts into some of our most productive agricultural land. Whether productivity continues or is only a temporary "bloom" depends on how well man is able to stabilize biogeochemical cycles and energy flow at the new increased rates. As the large volume of water passes through the irrigation system, salts may be left behind that will gradually accumulate over the years until they become limiting, unless means of avoiding this difficulty are devised. The water supply itself can fail if the watershed from which it comes is abused. The ruins of old irrigation systems, and civilizations they supported, in the deserts of the Old World warn that the desert does not continue to bloom for man unless he understands the laws of the ecosystem and acts accordingly.

TUNDRAS Between the forests to the south and the Arctic Ocean and polar icecaps to the north lies a circumpolar band of about 5 million acres of treeless country called the arctic tundra (Figure 7-7). Smaller, but ecologically similar, regions found above the tree limit on high mountains are called alpine tundras. As in deserts, a master physical factor rules these lands, but it is heat rather than water that is in short supply in terms of biological function. Precipitation is also low, but water as such is not limiting because of the very low evaporation rate. Thus, we might think of the tundra as an arctic desert, but it can best be described as a wet arctic grassland or a cold marsh that is frozen for a portion of the year.

Although the tundras are often known as the "barren grounds" and may be expected to have a relatively low biological productivity, a surprisingly large number of species have evolved remarkable adaptations to survive the cold. The thin vegetation mantle is composed of lichens, grasses, and sedges, which are among the hardiest of land plants. During the long daylight (long photoperiod) of the brief summer the primary production rate is remarkably good where topographic conditions are favorable (as in low-lying areas of Figure 7-7B). The thousands of shallow ponds, as well as the adjacent Arctic Ocean, provide additional food to tundra food chains. There is enough combined aquatic and terrestrial net production, in fact, to support not

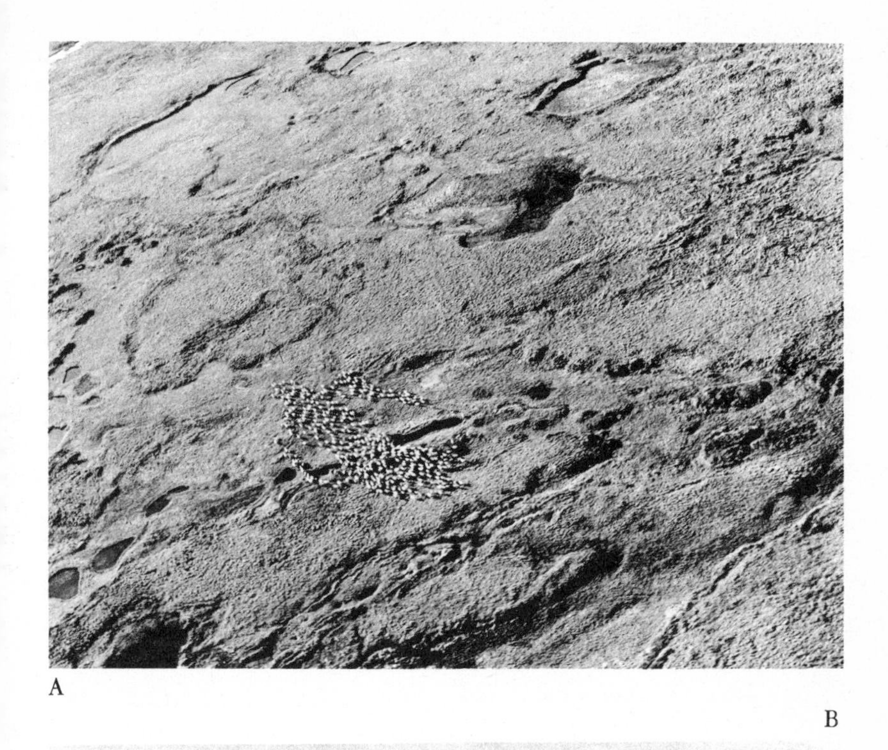

A

B

Fig. 7-7 The tundra. (A) Aerial view of tundra, showing herd of reindeer. The bumpy nature of the landscape is due to frost action; note, also, numerous small ponds. (B) Closeup of tundra in August near the Arctic Research Laboratory at Point Barrow, Alaska, showing grass and sedge vegetation. [(A) U. S. Department of Interior, Fish and Wildlife Service. (B) courtesy R. E. Shanks and John Koranda.]

only thousands of breeding migratory birds and emerging insects during the summer, but also permanent resident mammals that remain active throughout the year. The latter range from large animals such as musk ox, caribou, reindeer (Figure 7-7A), polar bears, wolves, and marine mammals, to lemmings that tunnel about in the vegetation mantle. The large land herbivores are highly migratory, since there is not enough net production in any one local area to support them. Where man tries to "fence in" these animals or select for domestication nonmigratory strains, such as domestic reindeer, overgrazing is almost inevitable unless judicious "rotating the pastures" is employed to offset the absence of migratory behavior. The dramatic ups and downs in the density of lemmings was discussed on page 127. The difference in response to light by plants in the arctic and alpine tundras was mentioned in Chapter 6. Man's impact on the tundra will increase as he strives to exploit oil and mineral resources from polar regions. Its special fragility needs to be recognized when roads and pipelines are built.

GRASSLANDS Natural grasslands occur where rainfall is intermediate between that of desert lands and forest lands (Figure 7-11). In the temperate zone this generally means an annual precipitation between 10 and 30 in., depending on temperature, seasonal distribution of the rainfall, and the water-holding capacity of the soil. Tropical grasslands may receive up to 60 in. concentrated in a wet season that alternates with a prolonged dry season. The IBP studies previously mentioned have shown that soil moisture is a key factor, especially as it limits microbial decomposition and recyling of nutrients. Large grassland areas occupy the interior of the North American and Euroasian continents. Other extensive natural grasslands are located in southern South America, central and southern Africa, and Australia.

Several aspects of North American grasslands are shown in Figure 7-7. Dominant plant-life forms are the grasses, which range from tall species (5 to 8 ft.) to short ones (6 in. or less) that may be bunch grass types (growing in clumps) or sod formers (with underground rhizomes). A well-developed grassland community contains species with different temperature adaptations, one group growing in the cool part of the season (spring and fall) and another in the hot part (summer); the grassland as a whole "compensates" for temperature, thus extending the period of primary production. The role of the C_3 and C_4 types of photosynthesis was discussed on page 77. Forbs (nongrassy herbs) are often important components, and woody plants (trees and shrubs) also occur in grasslands often in belts or groups along streams and rivers. Extensive areas in East Africa and other equatorial regions are occupied by a variant of the grassland biome, the *tropical savanna* where trees with interesting shapes are widely scattered in the grassland.

The grassland community builds an entirely different type of soil as compared to a forest, even when both start with the same parent mineral material. Since grass plants are short-lived as compared to trees, a large amount of organic matter is added to the soil. The first phase of decay is rapid, resulting in little litter but much humus; in other words, humification is rapid but mineralization is slow. Consequently, grassland soils may contain 5 to 10 times as much humus as forest soils. The dark grassland soils are among those best suited for growing man's principal food plants such as corn and wheat (Figure 7-8), which, of course, are species of cultivated grasses.

The role of fire in maintaining grassland vegetation in competition with woody vegetation in warm or moist regions was discussed in Chapter 4 (see Figure 4-11). Large herbivores are a characteristic feature of grasslands (Figure 7-8A). These are mostly large mammals, but large grazing birds are known to have occurred in the original fauna of New Zealand. The "ecological equivalence" of bison, antelope, and kangaroos in grasslands of different geographical regions was mentioned in Chapter 2. The large grazers come in two "life forms": running types, such as those mentioned above, and burrowing types, such as ground squirrels and gophers. When man uses grasslands as natural pastures he usually replaces the native grazers with his domestic kind—that is, cattle, sheep, and goats. Since grasslands are adapted to heavy enegry flow along the grazing food chain, such a switch is ecologically sound. However, man has had a persistent history of misuse of grassland resources by virtue of allowing overgrazing (Figure 7-8C) and overplowing. The result is that many grasslands are now man-made deserts. The importance of ecological indicators in the early detection of overgrazing was mentioned on page 112.

Morello's (1970) outstanding study of the interaction of fire and cattle grazing in Argentina has traced how large areas of Argentina's grasslands have become covered with thorny shrubs. Morello has proven that intensive cattle grazing reduces the combustible matter so that fires which are necessary to maintain grass cover can no longer burn. As a result, thorny shrubs formerly kept in check by periodic fires take over. The only way to restore grazing productivity is to expend fuel energy in mechanical removal and burning of woody vegetation. This is an example of a man-made vegetation change reversible only at great cost.

D

Fig. 7-8 Four aspects of grasslands. (A) Natural grassland with herd of bison on the National Bison Range in Montana. (B) Cattle grazing in natural grassland that is in good condition. (C) Overgrazed grassland that has the appearance of a man-made desert. (D) Grassland converted to intensive grain farming. [(A) and (D) U. S. Department of Interior, Fish and Wildlife Service. (B) and (C) U. S. Forest Service.]

What to do about the African grasslands that contain an unusual diversity of mammalian grazers is a question now facing the emerging nations of that area as they strive to raise nutritional levels in the human population. Many ecologists believe that it may be feasible to harvest the antelope, hippopotamuses, and wildebeests on a sustained-yield basis rather than exterminate them in order to substitute cattle. For one thing, the natural diversity means broader use of primary production. Further, the native species are immune to the many tropical parasites and diseases to which cattle are vulnerable.

FORESTS In Chapter 3 and again in Chapter 6, the point was made that the open sea and the forest are, in a comparative sense, the extreme natural ecosystem types in the biosphere in regard to standing crop biomass and the relative importance of allogenic and autogenic regulation. As shown in Figure 6-1, well-ordered and often lengthy ecological succession is characteristic with herbaceous plants often preceding trees. Consequently, in any one forest region we may see a mixture of vegetation including nonforest stages in succession as well as forest variants that are adapted to special soil and moisture conditions. Because the range of temperatures that will allow forest development is extremely wide (Figure 7-11), a sequence of forest types replace one another in a north-south gradient. Moisture is more critical to the tree than to the grass, but forests occupy a fairly wide gradient from dry to extremely wet situations.

Figure 7-9 shows three distinctly different forests in a north-south gradient. The northernmost forests, which form a belt just south of the tundra, are characterized by evergreen conifers of the genera *Picea* (spruce) and *Abies* (fir); species diversity is low, often with one or two species of trees forming pure stands. Deciduous forests are characteristic of the more southern moist-temperate regions; these forests have more pronounced stratification and a greater species diversity. Pines (*Pinus*) are found in both nothern coniferous and temperate deciduous forest regions, often as seral stages. The third type, the tropical forests, range from broad-leaved evergreen rain forests, where rainfall is abundant and distributed throughout the annual cycle, to tropical deciduous forests that lose their leaves during a dry season. Two life forms, the vine (lianas) and the epiphyte (air plants), are especially characteristic of tropical forests; a few species of these life forms are found in northern forests, but only in the tropical regions do they make up a conspicuous portion of the biological structure. Species diversity of both plants and animals tends to be high in tropical rain forests; there may be more species of plants and insects in a few acres of tropical rain forests than in the entire flora and fauna of Europe. Major differences in mineral cycling between tropical and temperate forests and their impact on agricultural conversion of forests were discussed in detail in Chapter 4. Jordan (1971) lists other interesting contrasts as follows: The ratio of leaf to new wood production is about 1:1 for the tropics and up to 1:6 for the temperate zone, which means tropical trees put proportionally more of their net production into leaves than into wood. Accordingly, annual leaf fall is greater in the tropics but the energy content in leaves is less per unit dry weight. Temperate deciduous forest leaves are often lobed and toothed while tropical leaves mostly have smooth edges. Following the

Fig. 7-9 Three forest types in a north-south temperature gradient. (A) Northern coniferous forest of spruce in Idaho. (B) Temperate deciduous forest of oaks, hickories, and other hardwoods in Indiana. (C) A tropical rain forest in Puerto Rico. [(A) and (B) U. S. Forest Service. (C) Courtesy University of Puerto Rico.]

A
B
C

A

B

terminology of natural selection outlined in Chapter 6, *r*-selection is favored in the temperate region and *k*-selection in tropics.

Two forest types in what might be thought of as a moisture gradient are shown in Figure 7-10. Chaparral occur in regions with winter rain and summer drought, and is a "fire-type" in that it is naturally subjected to fires and is adapted to this factor (see page 117). This kind of dwarf woodland is known as "macchie scrub" in the Mediterranean region and "Mallee scrub" in Australia. Other types of dry climate dwarf forests include the Pinon-Juniper of lower altitudes in the western mountains of the United States and the tropical thorn scrub forests of Africa. In contrast, the temperate rain forests, such as those along the coasts from northern California to Washington (Figure 7-10) occur where there is abundant moisture. They do not have as great a species diversity as tropical rain forests, but individual trees are larger and the total timber volume may be greater. The California redwoods are a variant of this forest type.

A good place in which to observe the pattern of forests in relation to climate and substrate is the Great Smoky Mountains National

Fig. 7-10 Three forests adjusted to different moisture and fire conditions. (A) Chaparral woodland, a dwarf forest of the winter rain-summer drought climate of coastal California; periodic fires are a major environmental factor. (B) A douglas fir stand in Washington, one of several forest types in the moist Pacific Northwest that develop some of the largest volumes of timber in the world. (C) A view of a leafy, two-storied tropical forest in Panama where there is a seasonal pulse in rainfall. The tall emergent trees with slender white trunks lose their leaves in the dry season while understory matrix of broad-leaved trees and palms are evergreen. As mentioned in the text, tropical forests tend to have a higher leaf to wood ratio and a greater annual production of leaves than do temperate forests. [(A) and (B) U. S. Forest Service. (C) Institute of Ecology, University of Georgia.]

C

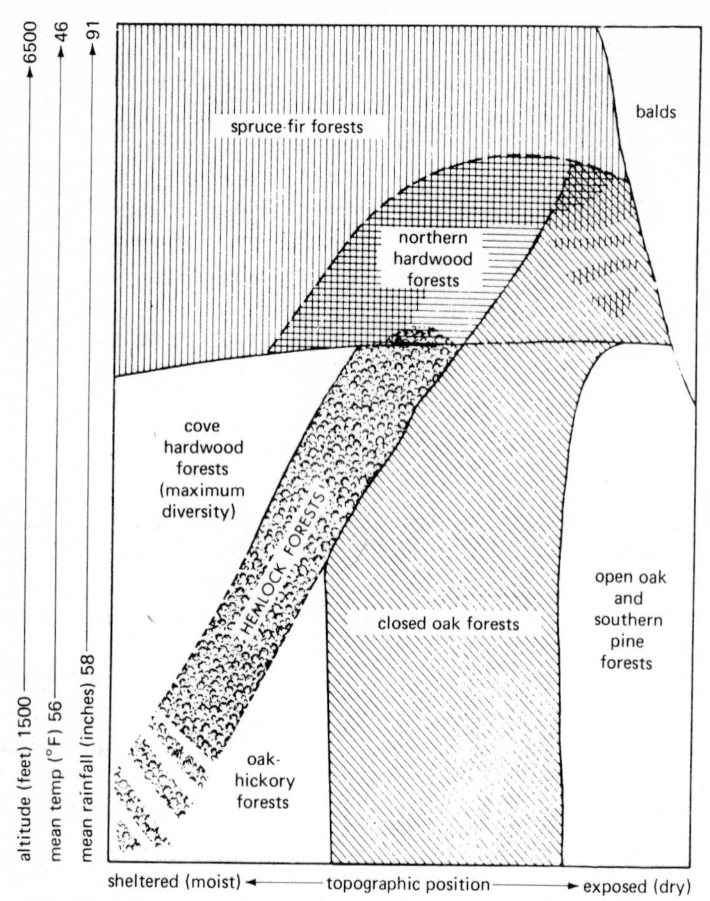

Fig. 7-11 The pattern of forest vegetation in the Great Smoky Mountain National Park as related to temperature and moisture gradients. See text for explanation. (Diagram prepared by R. Shanks. After R. H. Whittaker, *Ecological Monographs,* 1952.)

Park located along the Tennessee–North Carolina border. Figure 7-11 is a diagram that will help us view the landscape with the eyes of the ecologist. The altitude change produces a north-south temperature gradient, whereas the valley and ridge topography provides a gradient of soil moisture conditions at any given altitude. At sea level we would have to travel many hundreds of miles to observe the variety of climates present in a small geographical area in the Smokies. The pattern of vegetation along the gradients stands out best in May and early June (when floral displays are also spectacular), but the remarkable way in which forests adapt to topography and climate is evident at any time of year.

As shown in Figure 7-11, the forests of the Smokies range from open oak and southern pine stands on the drier, warmer slopes at low altitudes, to northern coniferous forests of spruce and fir on the cold, moist summits. The southern pine stands extend upward along the exposed ridges, and the northern hemlock forest extends downward in the protected ravines where moisture and local temperature conditions are those of higher altitudes. The maximum diversity of tree species occurs in sheltered (that is, moist) locations about midway in the temperature gradient.

The reason why some of the high, exposed slopes of the Smokies are covered with rhododendron thickets or grass instead of trees has not been adequately explained. These "balds" are not alpine tundras, for the altitude is not great enough for a true treeless zone. Whatever the reason (perhaps fire) for their original establishment, the shrub community is now so well established that it resists invasion by the forest. In this situation we can observe how whole communities, as well as the individuals in them, compete with one another. The eventual outcome may depend on the occurrence of some event such as fire or storms that might tip the balance in fa'or of one or the other ecological system.

Timber production and the practice of forestry pass through two phases. The first phase involves the harvest of net production that has been stored as wood over a period of years. When the accumulated growth of the past has been used up, man must adjust his forestry practice to harvesting no more than the annual growth if he expects to have any wood products at all. In the northwestern United States the first phase is still under way; the annual timber cut in this region is about double the annual growth. In contrast, the second phase has been reached in the southeastern United States. Most of the old timber has been cut; hence, forestry practice is primarily concerned with young forests where the harvest now balances annual growth. Although, as pointed out in Chapter 6, the annual net production in a young forest is often greater than that of an old forest, the quality of the wood for lumber use is not as good, since wood of fast-growing young trees is not as dense as that of slow-growing older trees. As in so many situations, the dichotomy between quantity and quality has to be recognized; rarely can we have both.

MAN'S FOREST EDGE HABITAT

Human civilization seems to reach the most intense development in what was originally forest and grassland especially in temperate regions. Consequently, most temperate forests and grasslands have been greatly modified from their primeval condition, but the basic nature of these ecosystems has by no means been changed. Man, in fact, tends to combine features of both grasslands and forests into a habitat for himself that might be called *forest edge*. When man settles in grassland regions he plants trees around his homes, towns, and farms, so that small patches of forest become dispersed in what may have been treeless country. Likewise, when man settles in the forest he replaces most of it with grasslands and croplands (since little human food can be obtained from a forest), but leaves patches of the original forest on farms and around residential areas. Many of the smaller plants and animals originally found in both forest and grassland are able to adapt and thrive in close association with man and his domestic or cultivated species. The American robin, for example, once a bird of the forest, has become so well adapted to the man-made forest edge that it has not only increased in numbers but has also extended its geographical range. Most forest birds in Europe have switched from the forest to gardens, cities, and hedgerows or else they have become extinct, since there are no longer many large tracts of unbroken forest. Most native species that persist in regions heavily settled by man become useful members of the forest-edge ecosystem of man, but a few become pests. The worst pests, however, are more likely to be species introduced from afar, as was discussed in Chapter 5.

If we consider croplands and pastures as modified grassland of early successional types, then man depends on grasslands for food, but likes to live and play in the shelter of the forest, from which he also garners useful wood products. At the risk of oversimplifying the situation we might say that man in common with other heterotrophs seeks two basic things from the landscape; "production" and "protection." But unlike lower organisms, he also finds aesthetic enjoyment in the beauty of natural landscapes. For mankind, forests provide all three needs, but especially the latter two. In many cases the monetary value of the wood, if harvested all at once, is far less than the value of the intact forest that provides recreation, watershed protection, home sites, and so on, plus a modest harvest of wood as well.

SUGGESTED READINGS

References cited

Cloud, Preston E. 1969. *Resources and Man.* San Francisco: W. H. Freeman.

Edmonson, W. T. 1968. Water-quality management and eutrophication: the Lake Washington case. In: *Water Resource Management and Public Policy,* ed. Campbell and Sylvester, pp. 139–178. Seattle: Univ. of

Washington Press. (See also layman's account in *Harper's Magazine*, June, 1967.)

Gasaway, Charles R. 1970. Changes in the fish population in Lake Francis Case in South Dakota in the first 16 years of impoundment. Technical Paper No. 56, Bureau Sport Fisheries and Wildlife, Washington.

Hammond, Allen L. 1972. Ecosystem analysis: biome approch to environmental research. *Science*. 175:46–48.

Heezen, B. C.; M. Tarp; and M. Ewing. 1959. The floors of the ocean. *Geol. Soc. Amer.* Special Paper 65.

Holeman, J. N. 1968. The sediment yield of major rivers of the world. *Water Res.* 4:737–747.

Jordan, Carl F. 1971. A world pattern in plant energetics. *Amer. Sci.* 59(4): 425–433.

Morello, Jorge. 1970. Modelo de relaciones entra pastizales y lenosas colonzadoras en el Chaca Argentino. (A model of relationship between grassland and woody colonizer plants in the Argentine Chaco.) *Idia.* No. 276, pp. 31–51. December 1970.

Schmidt-Nielsen, B. and K. Schmidt-Nielsen. 1964. *Desert Animals, Physiological Problems of Heat and Water.* New York: Oxford Univ. Press.

Wolman, M. G. 1971. The nation's rivers. *Science.* 174:905–918. (Excellent graphs and tables of water quality indices.)

The oceans

Baker, D. James, Jr. 1970. Models of oceanic circulation. *Sci. Amer.* 222 (1):114–121.

Berrill, N. J. 1966. *The Life of the Ocean.* (*Our Living World of Nature Series.*) New York: McGraw-Hill.

Carson, Rachel. 1952. *The Sea Around Us.* New York: Oxford Univ. Press.

Coker, R. E. 1947. *This Great and Wide Sea.* Chapel Hill: Univ. North Carolina Press.

Goldreich, Peter. 1972. Tides and the earth-moon system. *Sci. Amer.* 226 (4):42–52.

Hallam, A. 1972. Continental drift and the fossil record. *Sci. Amer.* 227(5): 56–66.

Hardy, A. C. *The Open Sea.* 1957, Vol. 1; *The World of Plankton.* 1959, Vol. 2; *Fish and Fisheries.* Boston: Houghton-Mifflin.

MacIntyre, Ferren. 1970. Why the sea is salt. *Sci. Amer.* 223(5):104–115.

Odum, E. P. 1971. *Fundamentals of Ecology*, 3rd ed. Chapter 12. Philadelphia: Saunders.

Sci. Amer. Special Issue on the Oceans. September 1969, Vol. 221, No. 3.

Thorson, Gunnar. 1971. *Life in the Sea.* New York: McGraw-Hill.

Wilson, J. T. 1963. Continental drift. *Sci. Amer.* 208 (4): 86–100.

Estuaries and seashores

Amos, William H. 1966. *The Life of the Seashore.* (*Our Living World of Nature Series.*) New York: McGraw-Hill.

Carson, Rachel. 1955. *The Edge of the Sea.* Boston: Houghton-Mifflin.

Dolan, R.; P. J. Godfrey; and W. E. Odum. 1973. Man's impact on the barrier islands of North Carolina. *Amer. Sci.* 61:152–162.

Lauff, George A., ed. 1967. Estuaries. *Amer. Assoc. Adv. Sci.*, Publ. No. 83. Washington, D.C. (an important reference work with many authors).

Odum, E. P. 1961. The role of tidal marshes in estuarine production. *The Conservationist.* June-July, pp. 12–15. New York Dept. Conservation, Albany.

_______. 1971. *Fundamentals of Ecology*, 3rd ed. Chapter 13. Philadelphia: Saunders.

Odum, William E. 1970. Insidious alteration of the estuarine environment. *Trans. Amer. Fish. Soc.* 99:836–847.

Stephenson, T. A. and Anne Stephenson. 1973. *Life Between Tidemarks on Rocky Shores.* San Francisco: W. H. Freeman.

Teal, J. and M. Teal. 1969. *Life and Death of the Salt Marsh.* Boston: Little, Brown.

Yonge, C. M. 1949. *The Seashore.* London: Collins. (New Naturalist Series).

Freshwater

Bennett, G. W. 1962. *Management of Artificial Lakes and Ponds.* New York: Reinhold.

Coker, R. E. 1954. *Streams, Lakes and Ponds*, Chapter 1 and Part III. Chapel Hill: Univ. North Carolina Press.

Deevey, Edward S. 1951. Life in the depths of a pond. *Sci. Amer.* 184(4): 68–72.

Eliassen, Rolf. 1952. Stream pollution. *Sci. Amer.* 186(3):17–21.

Fisher, Stuart G. and Gene E. Likens. 1972. Stream ecosystem: organic energy budget. *Bio-Sci.* 22:33–35.

Hall, Charles A. S. 1972. Migration and metabolism in a temperate stream ecosystem. *Ecol.* 53:586–604.

Hynes, H. B. N. 1962. *The Biology of Polluted Waters.* Liverpool, England: Liverpool Univ. Press.

Likens, Gene E. and F. H. Bormann. 1974. Linkages between terrestrial and aquatic ecosystems. *Bio-Sci.* 24:447–456.

Lowe-McConnell, R. H., ed. 1966. *Man-made Lakes.* New York: Academic Press.

Niering, William A. 1966. *The Life of the Marsh.* (*Our Living World of Nature Series.*) New York: McGraw-Hill.

Patrick, Ruth. 1970. Benthic streams communities. *Amer. Sci.* 58:546–549.

Ragotzkie, Robert A. 1974. The great lakes rediscovered. *Amer. Sci.* 62:454–464.

The terrestrial biomes

Allen, Durward L. 1967. *The Life of Prairies and Plains.* (*Our Living World of Nature Series.*) New York: McGraw-Hill.

Bell, Richard H. V. 1971. A grazing ecosystem in the Serengeti. *Sci. Amer.* 225(1):86–93.

Bourliere, F. and M. Hadley. 1970. The ecology of tropical savannas. *Ann. Rev. Ecol. and Systems.* 1:125–152.

Denison, William C. 1973. Life in tall trees. *Sci. Amer.* 228(6): 74–80.

Douglas, I. 1967. Man, vegetation and sediment yields of rivers. *Nature.* 215:925–928.

Hadley, Neil F. 1972. Desert species and adaptation. *Amer. Sci.* 60(3):338–347.

Hunt, Charles B. 1973. *Natural Regions of the United States and Canada.* San Francisco: W. H. Freeman.

Johnson, Phillip L. 1969. Arctic plants, ecosystems, and strategies. *Arctic.* 22:341–355.

Love, R. Merion. 1970. The rangelands of the western United States. *Sci. Amer.* 222(2):88–96.

McCormick, Jack. 1966. *The Life of the Forest.* (*Our World of Nature Series.*) New York: McGraw-Hill.

McGinnies, W. G.; B. J. Goldman; and P. Paylore, eds. 1969. *Deserts of the World.* Univ. Ariz. Press.

Noy-Meir, Imanuel. 1973. Desert ecosystems: environment and producers. *Ann. Rev. Ecol. and Systematics.* 4:25–51.

Polunin, N. 1960. *Introduction to Plant Geography.* New York: McGraw-Hill.

Richards, Paul W. 1973. The tropical rain forest. *Sci. Amer.* 229(6):58–67.

Riley, D. and A. Young. 1968. *World Vegetation.* England: Cambridge Univ. Press.

Shelford, V. E. 1963. *The Ecology of North America.* Urbana, Ill.: Univ. Illinois Press (Good basic description of biomes.)

Shelford, V. E. and S. Olson. 1935. Sere, climax, and influent animals with special reference to the transcontinental coniferous forest of North America. *Ecology.* 16:375–402. (A classic paper documenting how animals link together various developmental stages of vegetation within a major biome type.

Sutton, Ann and Myron. 1966. *The Life of the Desert.* (*Our Living World of Nature Series.*) New York: McGraw-Hill.

Walter, Heinrich. 1973. *Vegetation of the Earth, in Relation to Climate and the Eco-physiological Conditions.* Heidelberg and New York: Springer-Verlag.

chapter 8

Resources, Pollution, Bionomics, and Ecosystem Management

In the Preface and Chapter 1, the expanded scope of ecology as an academic subject, and the shift in the application of ecological principles from the component (population) to the ecosystem level were emphasized. General realization that the "supply depot" and the "living space" functions of one's environment are interrelated, mutually restrictive, and not unlimited in capacity has amounted to a historic "revolution in attitude" which is a promising sign that man may soon be ready to apply the principles of ecological control on a large scale. A change in public attitude toward the environment began first in the affluent countries, but is now slowly spreading to the less developed countries as political leaders begin to realize that it is in the best interests of each country, large and small, to be concerned with the big picture as well as with internal problems. In fact, it is obvious that internal problems dealing with such basics as food and energy cannot be solved on a small scale. As social critic Lewis Mumford (1967) has so well phrased it, "Ideological misconceptions have impelled us to promote the qualitative expansion of knowledge, power, productivity without inventing any adequate systems of control," and to forget that "quality in control of quantity is the great lesson of biological evolution." For more on the background of the new environmental awareness movement, see Lynn White's *The Historical Roots of Our Ecological Crisis* (1967), Lewis Moncrief's *The Cultural Basis for Our Environmental Crisis* (1970), Garrett Harden's *The Tragedy of the Commons* (1968), and my own essay, *The Attitude Revolution* (Odum, 1970).

All through this book we have cited examples of how ecological principles can contribute to achieving a mature balance between the systems of man and the systems of nature in such a way that quality controls quantity and human values are not sacrificed on the alter of technological advancement. We have repeatedly dealt with the shortcomings of the "one-problem/one-solution" or the "crisis" approach, and the need to model and act on a larger spatial scale and a longer time scale. The difficulties of enlarging the scale of decision-making are monumental because (1) society, its governments, and its educational and research institutions are excessively fragmented into numerous specialized "departments," (2) economic and political systems are excessively growth promoting and short-ranged in terms of goals, and (3) human behavior is such that public attention shifts rapidly from one crisis to another. Thus, we have first the "pollution crisis," then the "fuel crisis," the "food crisis," the "urban crisis," the "money crisis," and so on—or perhaps a repeat of the cycle all over again. The great challenge in public education is to keep the focus on the underlying problem as successive "crisis" are met and temporarily solved by some political or economic "quick-fix" maneuver. The ecologist holds that most of the current "crisis" are part of the continuing problem of how best to integrate man and nature.

In this chapter we shall try to apply the material presented in the preceding seven chapters to human affairs in a four-step sequence, as spelled out in the chapter heading. First to be considered are the "goods"—that is, the resources on which all depends. Next comes the "bads"—that is, the pollution disorder which requires attention in terms of energy and human effort if there is to be a net benefit in the use of resources. Next, we shall consider the concept of bionomics as a kind of expanded economics that includes cost-accounting of the works of nature as well as the works of man. Finally, we come to the ultimate goal, ecosystem management, that is, management of man and his environment as a whole rather than as separate entities. Underlying all of this is the need for a better understanding of the popula-

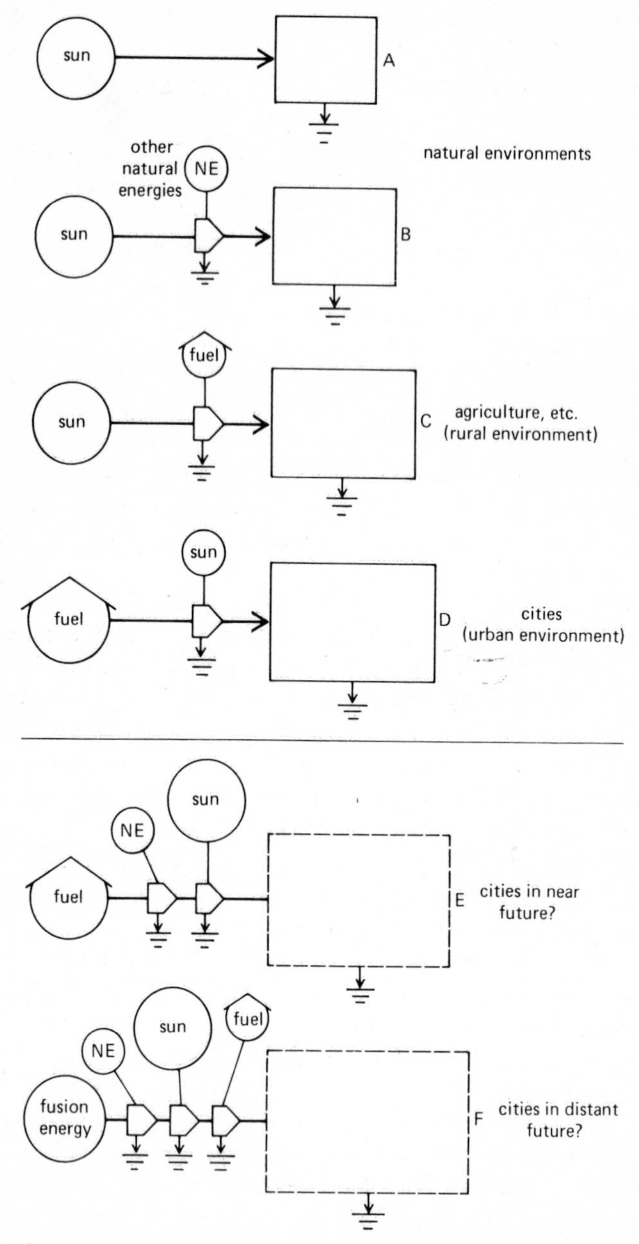

Fig. 8-1 Energy resources for the four basic kinds of ecosystems [(A)–(D)], and two possible energy scenarios for future cities [(E) and (F)].

tion-production complex on a larger scale—that is, how population growth and economic growth are related and how control of one affects the other.

ENERGY RESOURCES Figure 8-1 recapitulates Table 2-1 by showing in diagrammatic form the energy resources that power the basic kinds of ecosystems. In this diagram circles signify continuous sources of energy, that is, sources that may be expected to continue at the same level indefinitely. The "tank" symbol, on the other hand, indicates finite sources that decrease as used. This symbolic language was used and described in Chapters 2 and 3; see especially Figure 3-1. Figure 8-1A and B are the unsubsidized and subsidized natural ecosystems, respectively. The fuel-subsidized agricultural system is represented by diagram c and the fuel-powered city by d. Also shown are two possible energy scenarios for future cities. In the near future (e) more use will probably be made of sun and other natural energies so as to increase the efficiency and prolong the use of dwindling supplies of fuel. If nuclear fusion can be controlled at a favorable cost-benefit ratio, a more continuous supply would be available in the distant future (diagram f).

The energy resource situation in the United States, as well as in all of the other industrialized nations and most of the undeveloped ones can be very simply and bluntly stated as folows: Energy use is, or will very soon be, greater than that which can be supplied at a reasonable cost from sources within the boundaries of the nation. Which is to say that even where there are large reserves such as coal, oil shale, or offshore oil, the high cost of procurement and conversion will place severe constraints on economic growth and create difficult balance of trade problems for individual nations. Undeveloped countries that lack fuel resources are particularly hard hit when the price of fuel rises. The need to conserve energy (by reducing waste and increasing the efficiency of use), to allocate supplies on a worldwide basis, to increase efficiency of conversion of "difficult-to-get-at" sources, and to seek new sources will all receive the undivided attention of mankind for a long time into the future. This is a prime example of the long-range problem that cannot be solved on a crisis basis.

It is vitally important that everyone study carefully and strive to understand the nature of atomic energy, about which there is so much hope and controversy. It is especially important to distinguish between the several types of nuclear power. The kind of atomic energy now being used to generate electricity on a limited scale is based on the fission or "splitting" of the uranium with the release of energy, and also the release of dangerous "fission products," such as radioactive strontium and cesium. Some plutonium, an extremely dangerous radioactive substance (and, also, one that can be made into bombs), is also a by-product. Fission atomic energy is a "fuel" energy since the supply of fissionable uranium (^{235}U) is limited); actually there is less energy in this form than in coal left in the earth's crust. Also, tapping this source of energy is proving to be more troublesome and expensive than originally predicted. The breeder reactor now undergoing experimental tests, would prolong the uranium fuel supply since a more abundant form of uranium (^{238}U) can be used and new fissionable fuel is created in the reactor as the original fuel is used up. But an increased production of plutonium increases the radiation hazard.

Nuclear fusion is a different form of atomic energy entirely, one that involves the fusion of light atoms such as hydrogen to form a heavier atom with the release of energy. Extremely high temperatures are necessary for this energy release which resembles that which occurs on the sun. Fission products and plutonium would not be produced unless a fission reaction were used to create the temperatures necessary for fusion (as in the hydrogen bomb), but there would be problems with radioactive hydrogen (tritium). Controlling fusion involves containing the intense reaction, perhaps within magnetic fluxes or with laser beams, since no vessel could stand the temperature required. There is much discussion about hybrid fission-fusion systems, but any widespread use of fusion as a worldwide source of industrial energy is a long way in the future. Recent issues (1972, 1973, and so on) of *Science and Public Affairs: Bulletin of the Atomic Scientists* contain many informative and largely nontechnical articles on nuclear energy. For a nontechnical book, see Inglis (1973).

We have already discussed in some detail solar energy (see page 17). This abundant, but dilute and low quality energy resource can be put to work in cities doing low level "jobs," such as heating water, commercial buildings, and dwellings—thus sparing fuel for other uses. Extensive use of solar energy in the place of fuel requires a technology not yet developed. Direct conversion of sunlight into electricity by means of solar cells is a promising new technology now under intensive study. Another way we could upgrade solar energy for higher level work would be to make use of nature's efficient conversion, namely photosynthesis, for fuel as well as for food. Szego (1971) has calculated that the annual growth of wood in managed "fuel forests" could supply the United States with substantial amounts of electricity if burned in wood-fired, steam-electric plants. This would be a "circle" source rather than a "tank" source, as symbols are used in Figure 8-1. The long-term cost-benefit of such a use is yet to be calculated but it is a possibility worth considering in regions where forests are climax, human population density low, or where there is a lot of hilly land not suitable for agriculture or other nonforest uses.

FOOD AND FIBER RESOURCES

The "food for man" situation was discussed in detail in Chapter 3. Figure 8-2 reemphasizes the key role that energy subsidies play in the production of food and fiber (cotton, wool, paper, wood, and so on). The age-old paddy rice culture is very efficient in terms of food yield per unit of energy subsidy, but it is backbreaking for the people who plant and harvest the rice. At the other extreme, feedlot

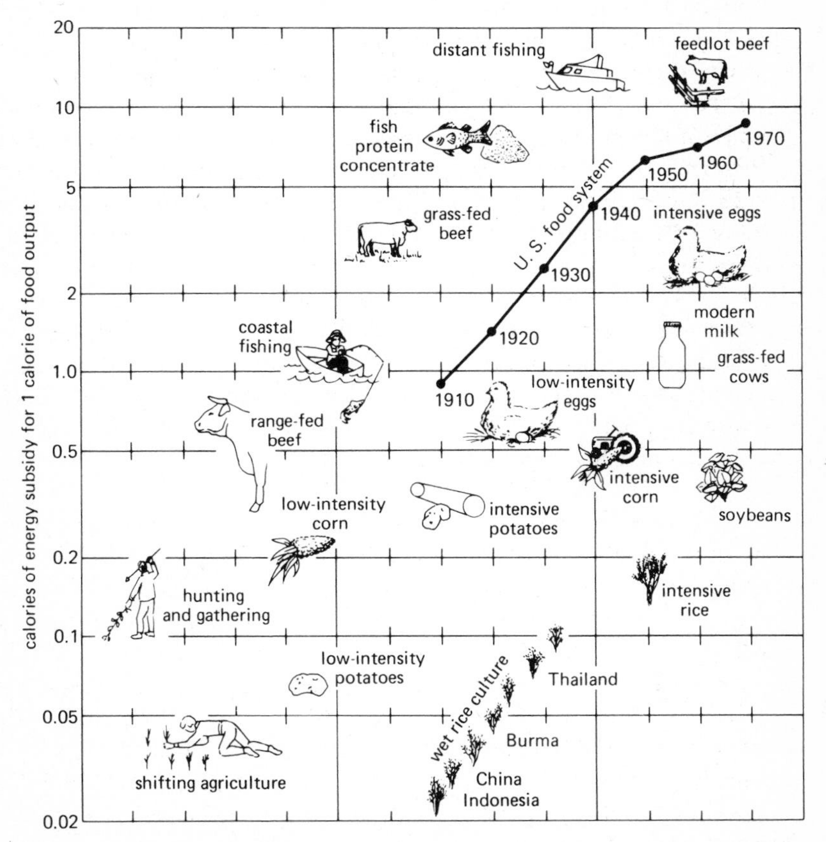

Fig. 8-2 Energy subsidies for various food crops. The energy history of the U. S. food system is shown for comparison. (After Steinhart and Steinhart, *Science*, Vol. 184, 1974. Copyright © 1974 by The American Association for the Advancement of Science.

beef requires 10 cal of fuel energy for every calorie of food produced, but neither man nor beast has to do much work. Feedlots do not make very good ecological sense for another reason. Cows have a marvelous adaptation—the rumen, which enables them to convert very low protein food such as grass and hay into high protein food. When cows are fed rich grains in a feedlot, this adaptation is bypassed, and the meat produced tends to be too fatty for good human health. Also, feedlots produce severe watershed pollution that adds another stress on the environment, and another cost for man. There is much to be said for putting the cow back on grass.

Avoidance of the boom-and-bust syndrome, as discussed on page 125 is another reason for considering a somewhat less energy-intensive agriculture (for example, crop systems in the midsection of Figure 8-2), especially for undeveloped countries. It is difficult and costly in terms of energy to sustain very high yields of the same crop over long periods of time. In Figure 8-3 the history of cotton yields in a valley in Peru is graphed. Prior to 1958 moderately intensive culture resulted in moderate yields until the bad year of 1949. Then a highly industrialized monoculture was put into effect including mass aerial spraying of chlorinated hydrocarbon pesticides (the DDT family of chemicals). Yields per acre were increased (and also total yields as more land was devoted to cotton), but the cost of chemical control also increased as insects became immune, and new pests apeared to replace the old ones (cycle 2, in Figure 8-3). In fact, the number of

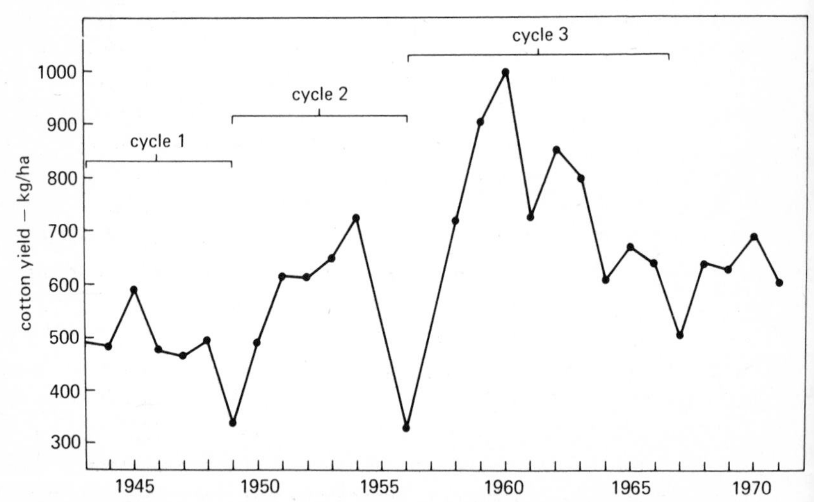

Fig. 8-3 Three cycles in yield of cotton in the Cañete Valley of Peru associated with three successive strategies for insect control. See text for explanation. (Data for 1943–1963 from T. B. Barducci [1969: Table 23-2]; data for 1964–1971, personal communication with Dr. Barducci.)

major insect pests increased from 8 to 13 species. This system collapsed completely in 1956, and other crops were grown instead of cotton. During the next several years (cycle 3, in Figure 8-3), a diversified agricultural program evolved with less intensive cotton culture and what entomologists now call "integrated control" of insects, that is, a mixture of biological and chemical controls specifically tailored for more pinpoint control of particular species. (also avoiding the environmental pollution caused by mass spraying of nonspecific poisons). Yields per acre increased markedly for several years, but by 1964 this boom had tapered off. From then on average yields were not too much better than it was in the 1940s. This well-documented situation may be an example of the principle discussed in Chapter 4, namely, that intensive monoculture of annual plants of the C_3 type is less appropriate for warm latitudes than for cold latitudes.

Forest and wildlife management faces the same dilemma as encountered in agriculture, namely, how far to go in reducing diversity in order to increase yield. To what extent, for example, are monocultures appropriate for forestry? A possible "trade-off" policy for our National Forests was briefly discussed on page 164.

MINERAL RESOURCES

All the energy in the world would be of no avail if the nutrients required by life and the materials required by commerce are not continuously fed into the ecosystem, or recycled within the ecosystem, or both. The interdependence of energy flow and material cycling is shown in Figure 4-1 and the basic principles of nutrient cycling in the biosphere were discussed in Chapter 4. The point was also made that as man depletes the reservoir storage bins, recycling not only becomes necessary, but some of the energy flow has to be diverted from productive processes (that is, new growth) to power the recycle process. Figure 8-4 provides an overview of three alternate depletion patterns for minerals such as iron, copper, aluminum, and so on, required in huge quantities by industrialized civilization. A one-way pattern of unrestricted mining, use, and throwaway is projected to lead to a boom-and-bust, as shown by curve a. Some key metals, such as copper, could be "mined out" by the year 2000. Depletion time could be extended by partial recycling and less wasteful use, as shown by curve b. Efficient recycle, including the necessary energy allocation, combined with stringent conservation and substitutions (switching to a more abundant material whenever possible) can extend the mineral depletion curve substantially (curve c). It remains to be seen whether mankind will, or can, make a choice between these options, or whether events will

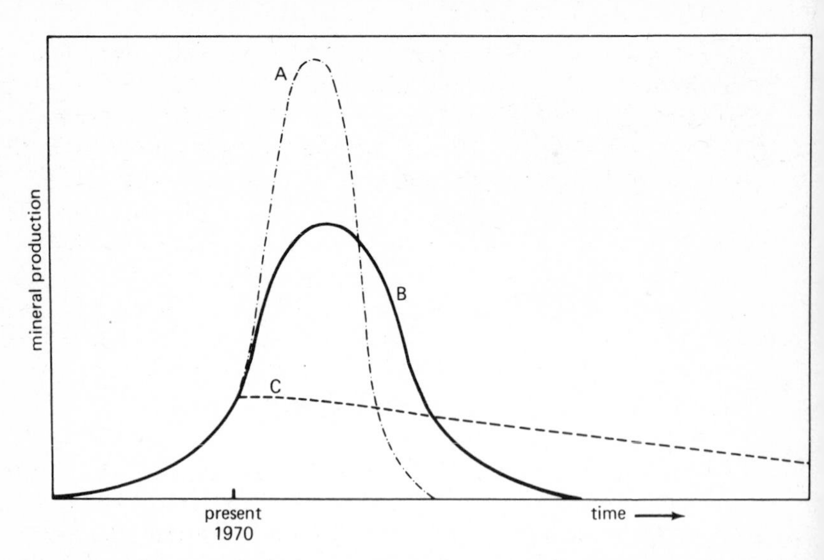

Fig. 8-4 Alternate depletion patterns for mineral resources. (A) A huge "boom and bust" (rapid increase in extraction rate then decline to zero) is predicted if present custom of unrestricted mining, use, and throwaway continues. (B) Depletion time can be extended by partial recycle and less wasteful use. (C) Efficient recycle combined with stringent conservation and substitution can extend mineral depletion curves indefinitely (After Cloud, 1969, from Odum, *Fundamentals of Ecology*, 3rd ed. W. B. Saunders, 1971.)

force some other choice. As with fossil-fuel energy the total amount of minerals in the earth's crust (and in sea water) is very large, but as the concentrated, easily mined supplies are dispersed, procurement becomes increasingly costly both in a monetary sense and in terms of ill effects on the environment. For a comprehensive review see the National Academy's report, *Resources and Man*, edited by Cloud (1969).

POLLUTION DISORDER The word pollution is derived from a Greek root meaning "defilement." The National Academy's comprehensive report-review *Waste Management and Control* (Spilhaus, ed., 1966) defines pollution as an undesirable change in the physical, chemical, or biological characteristics of air, water, or land that will be, or may be, harmful to human and other life, industrial processes, living conditions, and cultural assets. In Chapter 3 we considered pollution in the broad sense as the thermodynamic disorder that is the by-product of energy conversion and the use of resources. Pollutants, then, are the "bads" that detract from, and potentially limit, the use of the "goods" (that is, resources).

Pollutants are produced by natural ecosystems as well as by man's agricultural and industrial activity. However, nature by and large "treats" (that is, renders less harmful), recycles, or makes good use of her pollutants. In the past man has counted on nature to treat his pollution as well. As the twentieth century comes to a close the sheer volume and increasingly poisonous nature of man-made pollutants threatens the integrity of nature and the cultural development of man. There is no way to avoid pollution entirely (since we cannot circumvent the Second Law of Thermodynamics), but there are many ways to curtail the amount and to reduce the harmful impacts. Pollution control, then, must receive the same high priority in human affairs as energy conversion because they are linked.

Recognizing that there are several distinctly different kinds or basic types of pollution may be helpful when it comes to thinking about the overall problem of pollution abatement. First, are the *biodegradable pollutants*, such as domestic sewage, that can be rapidly decomposed by natural processes, or in engineered systems, such as municipal sewage treatment plants, where electrical energy and machines are used to speed up natural decomposition by microorganisms. In other words, this category includes substances for which there exists efficient natural waste treatment mechanisms that have evolved through eons of time. Heat (sometimes called thermal pollution), carbon dioxide, nitrates, and other by-products of metabolism and the complete combustion of fuels are examples since these are readily disposable by natural means. Problems arise with degradable types when input into the environment exceeds the decomposition, dispersal, or recycling capacities. Severe problems with sewage and other organic wastes in cities result from the fact that they have grown faster than treatment facilities. The technology for the treatment of domestic wastes is well developed; it just takes money and a greater public awareness that the added cost for waste treatment is one of the necessary prices that has to be paid for concentrated urban life.

It is customary to consider the treatment of degradable wastes in three stages: (1) *primary treatment*, a mechanical screening and sedimentation of solids (which are burned, buried, or hopefully in the future processed for fertilizer); (2) *secondary treatment*, a biological reduction of organic matter, as noted above; and (3) *tertiary or advanced treatment*, the chemical removal of phosphates, nitrates, persistent organics, and other materials. If we designate the cost of primary treatment as "1," then secondary treatment is 2–3 times as expensive, tertiary treatment to remove nutrients 5 times and tertiary treatment to produce drinkable (recycled) water 10 times. The reason that tertiary treatment is so much more expensive than secondary treatment is that a lot of special equipment and fuel energy are needed for the chemical removal stage; in contrast, the natural microorganisms do most of the work in a secondary treatment plant. The goal of most cities is to achieve 100 percent secondary treatment with the expectation that nature will be able to handle most or all of the tertiary treatment (see page 21).

Nondegradable pollutants constitute a second major class that include aluminum cans, long-chain detergents, glass, phenolic chemicals, plastics, and hundreds of man-made materials that do not degrade, or degrade very slowly, in the natural environment. In other words, these are substances for which there are no natural decay or treatment processes that can keep up with the rate of man-made input into the environment. Nondegradable products are a part of the "solid wastes" of cities that often end up in landfills. More and more in the future attempts will be made to separate degradable and nondegradable materials in the solid waste stream, with the former used as fuel and fertilizer and the latter recycled. Recovery of metals and plastics will become more feasible as the scarcity of new materials makes it worthwhile to expend the energy necessary for recycling. Another approach is to substitute degradable or easily recyclable materials for those that are neither. It has repeatedly been shown that the energy cost of the aluminum throwaway can is greater than that of most other containers, yet they continue to be made in ever-increasing quantities. In this case laws or regulations backed by strong public opinion would be necessary to bring about substitution because there is a short-term economic gain in the manufacture of the aluminum throwaway can.

Finally, we come to the third class of pollutants, *the poisons*, including such items as salts of heavy metals (mercury, lead, cadmium, and so on), smog gasses (see page 98), radioactive substances, pesticides, and an increasing array of industrial and agricultural chemicals whose toxicity to man and other life is incompletely known. Many of the poisonous pollutants have the tendency to concentrate in the food chain, as described on page 103. Whether degradable or not these pollutants interfere with vital bioenvironmental processes and pose a direct threat to human health. There is evidence, for example, of a link between air and water pollution and recent increases in human sickness and death from lung diseases and certain malignant neoplasms. Lave and Seskin (1970) estimate that a 50 percent reduction in air pollution in U.S. cities would save 2 billion dollars annually in aggregate cost of medical care and work hours lost in pollution-related sickness and disability. Dealing with hazardous pollutants is a problem because it is difficult to prove in a court of law who is responsible for a given pollutant, or that any one substance as present in low concentration in the environment is damaging to human health. Society as yet has no procedure for dealing with the synergistic effect of numerous low-level poisons that singly have little effect, but which together become deleterious.

The history of man's use of chemicals to combat insect pests provide, perhaps, an example of an evolving strategy for situations where it is not possible or feasible simply to stop producing the poisons altogether. To establish an objective overview of the highly controversial subject of pest control it is helpful to think in terms of what Carroll Williams (1967) has called the "three generations of pesticides," namely, (1) the botanicals and inorganic salts (arsenicals, and so on); (2) the DDT generation of organic, broad-spectrum poisons; and (3) the hormones and other narrow-spectrum biochemicals together with biological controls (parasites, and so on) which aim at pinpoint control without poisoning the whole ecosystem. These three generations more or less parallel the three cycles in the Peruvian cotton story, as diagrammed in Figure 8-3.

The first generation of pesticides were adequate to keep grandfather well fed when farms were small and diversified, farm labor plentiful, and cultural practices favorable to block massive build-up of pest populations. There was little accumulation of poisons beyond the crop field. The second generation of pesticides, the organoclorines (such as DDT) and organophosphates, ushered in an era of industrialized agriculture involving larger farms, more monoculture, substitution of fuel for human labor, and the genetic selection of crops adapted to massive fuel and chemical subsidies (see page 00). During this period poisonous chemicals spread far beyond the crop fields. Unfortunately, excessive optimism that the DDT generation had forever solved insect problems and freed the world from the threat of starvation lulled agricultural scientists into neglecting other approaches to insect control. The ability of insects to develop immunity to the new poisons and the damage done by the killing of useful parasites and predators was grossly underestimated. There followed an almost frantic escalation in the spraying of these potent poisons as the "dose" required to save the crop got bigger and bigger. This amounted to "overkill" as was dramatically brought to public attention in 1962 by Rachel Carson's famous book, *Silent Spring*. Finally, by the early 1970s pesticide pollution became so bad as to directly threaten human health with the result that some of the more persistent (that is, slow to degrade in the environment) organochlorines had to be outlawed. Interest then turned to the possibilities of the third generation of chemicals and to that previously mentioned (page 00) strategy of "integrated control" that involves a judicious mixing of degradable chemicals, biological control, diversified culture, and genetic selection for resistance. Among the promising narrow-spectrum chemicals are the species-specific juvenile hormones that prevent maturation and breeding, and various attractants that lure the target species to traps or poisons. One thing is certain: man's competition with insects for food will continually require vigilance and changes in strategy, with reasonable control and coexistence a more attainable goal than complete elimination of pests. Dr. Robert van den Bosch, a highly regarded entomologist, has recently stated that "in all of entomological history no broadly adapted insect, widely established over diverse terrain, has been eradicated through human effort" (*Science*, 184:112).

BIONOMICS The word bionomics means literally "management of life" and is derived from the same root (nomic: management) as economics, the term that literally means "management of the house." In most dictionaries bionomics is listed as a synonym of ecology, but the word may now be appropriate for an expanded economics of the ecosystem in which monetary values, cost accounting, and management of natural process are included along with man's works. Such a special word might not be needed if economists were enthusiastic about extending their traditional discipline to include the works of nature. Most economists express the opinion that there is more than they can cope with in man's "house" and that another discipline is needed to deal with the combined "house" of man and nature. In any event, all agree that the time has come to put greater emphasis on the value of the work of natural systems and to the impacts, both good and bad, that are external to business operations.

Table 8-1 provides an example of how monetary evaluation of a natural system might be extended to include useful work performed by self-maintaining ecosystems. Lines 1 and 2 represent conventional

Table 8-1. Four Bases for Economic Evaluation of a Tidal Estuary

Basis for Evaluation	*Annual Return per Acre*	*Income-Capitalization Value per Acre (at Interest Rate 5%)*
(1) Commercial and sport fisheries	$ 100	$ 2000
(2) Aquaculture potential	350	7000
(3) Tertiary waste treatment capacity[a]	2500	50,000
(4) Total life-support value[a]	4100	82,000

[a] See text for explanation of calculations. Data in the table from Gosselink, Odum, and Pope (1974).

bases for economic evaluation, while lines 3 and 4 extend the valuation to include the capacity of the tidal estuary to assimilate wastes and to provide general life support for man's fuel-powered systems. Two values are shown for each line, an annual return and an income-capitalized value obtained by dividing the annual return by an interest rate (5% or 0.05 in this case), a standard procedure in resource economics (see Barlowe, 1965).

Although natural seafood harvest and recreational values accruing from very large areas are impressive, they are small on an acre basis when compared with real estate values that the estuary might have if converted from its natural state to some developed state (as, for example, if the estuary was filled in for housing or factory development). Oyster culture or other intensive aquaculture would increase somewhat the commercial return from the estuary. However, the estuary in its natural state has a much greater value to the public as a whole in terms of its waste assimilation capacity and general life support, especially as the intensity of adjacent man-made development increases. The waste assimilation value, as estimated in Table 8-1, is based on the cost of tertiary treatment in treatment plants built and maintained by man. In other words, this is what it would cost society to treat wastes in amounts not exceeding the reasonable capacity of the estuary to metabolize treated municipal and nontoxic industrial wastes if the estuary was not available to do this useful work. The estimate for general life support was calculated by multiplying the total productivity of the estuary times an energy-dollars conversion factor. One such conversion suggested by H. T. Odum (1971) is based on the ratio of the gross energy consumption and the gross national product (GNP) for the country or region in question. As would be expected, the energy/GNP ratio varies in different countries. In an energy-conservative country such as New Zealand, that does not have a lot of heavy industry, about 7000 kcal of energy is consumed annually for each dollar of GNP. For the United States the energy-dollar ratio has fluctuated between 10,000 and 25,000 kcal per dollar between 1945 and 1970. Using a conservative figure of 10,000 kcal = one dollar and annual production rate for the estuary of 10,000 kcal m^{-2} or 41×10^6 per acre, then the annual return comes to \$4100 and the long-term value of an acre amounts to \$82,000 (see line 4, Table 8-1). Since "productivity" is a measure of a natural system's capacity to do all kinds of useful work, such as waste treatment, CO_2 absorption, O_2 production, seafood production, wildlife habitat maintenance, protecting cities from storms, transportation, and so on, then converting work energy to money is a convenient way of making a bionomic evaluation of a given natural system.

In the example just given what might be called the "social values" (lines 3 and 4, Table 8-1) inherent in preserving a natural estuary exceed the immediate or short-term commercial values. Unless the former values are recognized and appropriate action taken to preserve them, the pricing system based on incomplete accounting will tend to force an irreversible artificial development of estuaries, floodplains, watersheds, and prime farmlands even though it is in the general public interest that such areas continue to function as life-support ecosystems. The water situation briefly mentioned on page 103 is another example of how an urban dweller benefits from nature's recycling work. The cost of water produced by a natural watershed is very much less than the cost of artificially recycled water.

Putting a monetary value on the "free work of nature" does not solve the problem of the conflict of interest between the value to the property owner and the value to society where the area in question has a high social or public value in its natural state, but also a high real estate value if developed to something else. But at least this approach helps bring general recognition of values that are either not recognized or in danger of being lost through public apathy. For more on this approach see Gosselink, Odum, and Pope (1974).

As alluded to in the discussion of pollution, environmental debits as well as assets need to be included in economic assessments, especially environment-damaging residuals of manufacturing which in all but a very few countries escape control by the economy. A variety of incentives, sanctions, laws and governmental controls aimed at closing this dangerous gap in economic institutions are being discussed and tried out. Evidences of citizen concern are the increasing number of lawsuits aimed at halting pollution and what individuals and groups consider unwise alterations of the environment. In turn this has stimulated a judicial interest in cases involving the environment. Environmental law has become a new research and teaching focus in many law schools. From these largely uncoordinated efforts some generally agreed upon ground rules will hopefully emerge. There is no shortage of good ideas for economic and judicial reforms if the number of new books and articles on the subject is any indication. A sample of these are listed in the "Suggested Readings" list at the end of the book.

In the United States the National Environmental Protection Act (NEPA) represents the first attempt to provide a nation-wide legal basis for extending value systems to include the natural environment. The Act requires that "impact statements" be prepared for all large proposed man-made alterations. Hopefully this admitted stop-gap measure will lead to a total assessment procedure that includes environmental and social cost-benefits along with purely economic ones.

Perhaps the ultimate solution to the problem of joining ecological and economic values is to adopt energy units instead of monetary units for all values. The value of goods and services can certainly be measured in energy units as well as in dollars and cents; and, as we have seen, the value of the work of nature can best be expressed in energy units. Perhaps energy will prove to be the basic "currency" for the proposed new science of bionomics.

CONTROL THEORY At this point it would be well to review the elementary concepts of *cybernetics*, which is the science of controls. Control in any system, whether a simple temperature-regulating home-heating system or a complex ecosystem, depends on *feedback*, which occurs when output (or part of it) feeds back as input. Referring back to Figure 1-3, if we were to draw a curving line from the output arrow back to the input arrow, we would have added a "feedback loop" to the system diagram. When the feedback input is positive (like compound interest, which is allowed to become part of the principal), the quantity grows. *Positive feedback* is thus "deviation accelerating" and, of course, is necessary for the growth and survival of young organisms and young ecosystems, as pointed out in chapter 6. However, to achieve an orderly growth that does not "boom and bust" there must also be *negative feedback*, or "deviation-countering" input, especially as limits are approached.

In studying natural ecosystems we are impressed with the numerous and intricate negative feedback mechanisms that have evolved, as, for example, the controls on foliage-eating insects in forests, as described in Chapter 1. Man's fuel-powered urban and agricultural systems have been, until recently at least, under the influence of strong positive feedback with economic and political policies providing strong forcing functions for a quantitative growth. Many people have assumed that negative feedback will automatically develop as saturation is approached, and, therefore, it is not necessary for us to plan very far ahead or worry about the future. Such an attitude is completely untenable when growth is rapid; negative feedback must be established *well before* limits are reached to prevent dangerous overshoots (recall our discussion of exponential growth on page 125).

ECOSYSTEM MANAGEMENT Now let us move from the general consideration of control theory to specific examples to show that people's attitudes toward the environment are really changing. Two examples from California should suffice to illustrate trends. First, the matter of deciding where future atomic power plants should be placed. In terms of environmental costs the least expensive site would be along the coast where large volumes of seawater are available for cooling purposes. Inland location with freshwater cooling ponds would be the next cheapest, while installation of wet cooling towers would increase costs. The most expensive option would be dry cooling towers, which, however, would spare water for other uses. Ten years ago decisions would have been made only on economic and engineering grounds with the likelihood that many power plants would be located along the coast. Now, with the public and environmentalists deeply involved in the discussions and decision-making, it seems likely that very few power plants will be built on the coast because the natural scenic and recreational values are being given greater value, as evidenced by recent state wide vote in favor of the coastal protection "proposition 20." The combination of setting aside large areas as natural reserves and the increasing cost of energy can act as powerful negative feedback that could be a beneficial factor in slowing urban-type growth to a manageable rate.

A remarkable little book entitled *The California Tomorrow Plan*, edited by Heller and first published in 1971, is evidence of a reorientation of values and priorities on an even larger scale. This report, prepared by a citizens' group, assisted by various professional groups, proposes that a state commission be set up by the legislature to design a land-use plan for public debate and eventual adoption. Such a plan would regulate the kind and intensity of development and would identify and preserve the "life-support" function of the natural environment (as discussed on page 17), as well as preserve the quality of life in the cities. A unique feature of the plan is the proposed setting aside of large "reserve" areas where decisions on use would be postponed for sometime in the future, thus providing flexibility for future contingencies. Widespread resistance to planning at the present time is based on the very real fear of loss of individual freedoms as a result of increasingly dictatorial and bureaucratic control. As shown in Figure 8-5, which is reproduced from Heller's book, this need not be the case. Regional control of environmental matters would seem to be a good compromise between central control which would not be responsive to local needs, and purely local control (as is now the case), which is inadequate to cope with the scale of the problems. In other words, what seems to be needed *is expanded control of vital environmental matters in order to reduce central control of personal matters.* Freedom is not the absence of constraints, but is measured in terms of a variety of options available to the individual. Preservation of a wide variety of environments in functional balance provides the basis for

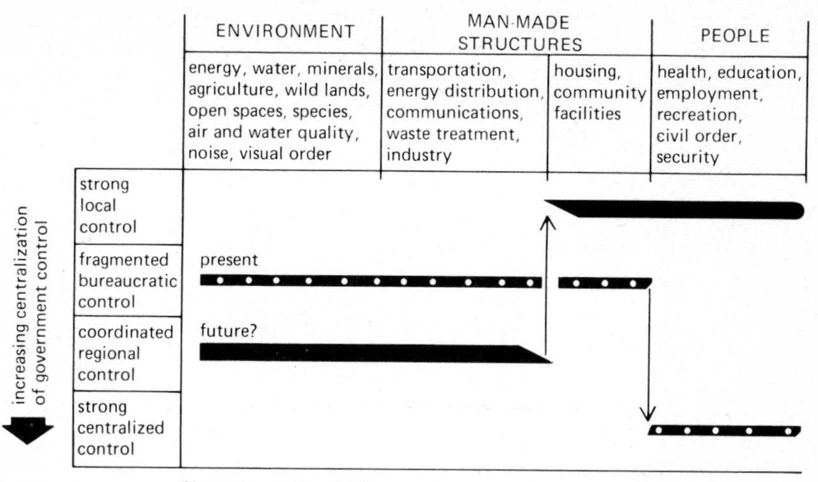

Fig. 8-5 Patterns of governmental control with respect to environment, man-made structures, and people. Future adoption of comprehensive regional land-use plans would greatly improve management of the environment, transportation, energy distribution, etc. At the same time facilitating the return of schools, housing, and civil order to local control, thus reversing the present trend of increasingly oppressive and inefficient central control. (Diagram after Heller, 1972.)

choices. This, then, must be the aim of "ecosystem management," a new venture for mankind. There has been a great surge of interest in research on comprehensive planning accompanied by growth in student enrollment in University Schools of Environmental Design, Environmental Engineering, Natural Resources, Urban Planning, Schools and Departments of Social and Political Science, and the like. Foundations, "think tanks," and consultant groups are all responding to the need. Perhaps soon Schools, Institutes, and Centers of Ecosystems Management will be formed to coordinate all of these special interest applied groups.

The ground work for comprehensive planning was laid by social scientists many years ago, at least in terms of ideas and recognition of needs. As public interest increases physical and biological scientists, engineers, and other technologists, and, somewhat reluctantly, political leaders are bringing their expertise to bear on implementing these ideas "whose time has now come." Especially important is the concept of "regionalism," as developed by the late Howard W. Odum and his colleagues and outlined in the books *Southern Regions* (1963) and *American Regionalism* (1938). Regionalism, as an approach to the study of society, is based on the recognition of distinct differences in both cultural and natural attributes of different areas which, nevertheless, are interdependent. Therefore, thorough inventories of man, resources, and natural environment provide the basis for coordination of regions that otherwise tend to engage in too much "competition exclusion" among themselves. Regional soil conservation of planning, which began in the 1930s, is an example of a highly successful activity that, in general, has maintained a good balance between local control and self-interest, and regional and national goals. As an aside we should point out that in recent years this program has tended to become too bureaucratic and centralized, a trend that must be guarded against, as noted in the preceding paragraph. At first, interest in regional planning was motivated by the desire to upgrade economically depressed regions. But now it is the economically affluent regions that most need comprehensive planning. Thus, the real goals of regionalism, namely, integration of regions and integration of man and nature now come to the forefront. The social science concept of different cultural units functioning together as a whole is, of course, parallel to the ecologist's concept of "ecosystem."

To reiterate, the biggest stumbling blocks to regional planning and ecosystem management are (1) the fragmented professional and political mechanism for decision-making, as just described; and (2) the fact that it is extremely difficult to determine the "human carrying capacity" of a region or the optimum level of population density and industrial development. In addition to available energy the manner in which the different ecosystems (as diagrammed in Figure 8-1) are coupled, and the proportional space occupied by each are important considerations. In Figure 8-6A ecosystems are reduced to two types: (1) natural systems powered by natural energies; and (2) man's fuel-powered and fuel-assisted systems, in order to illustrate one approach to the problem of determining optimum mixes. The value of the natural and developed systems, as might be calculated separately, interact to produce a third value shown by the "diversity interaction" box in Figure 8-6A. This, in effect, represents the total value or "well-being," as might be measured in net-benefit energy units, of the integrated system of man and nature. The developed system places a large stress on the natural system, as shown by the stress heat-sink arrow in Figure 8-6A. As the size and power of the developed system increases the natural system is degraded until a point is reached where the developed system begins to suffer from lack of life support; then the total value (value 3) declines. If equations are written for the energy circuits, and a computer program set up so that the proportion of developed to undeveloped land can be varied (see "switch" in Figure 8-6), then *performance curves* can be generated to show the relationship between the three values and the intensity of develop-

A

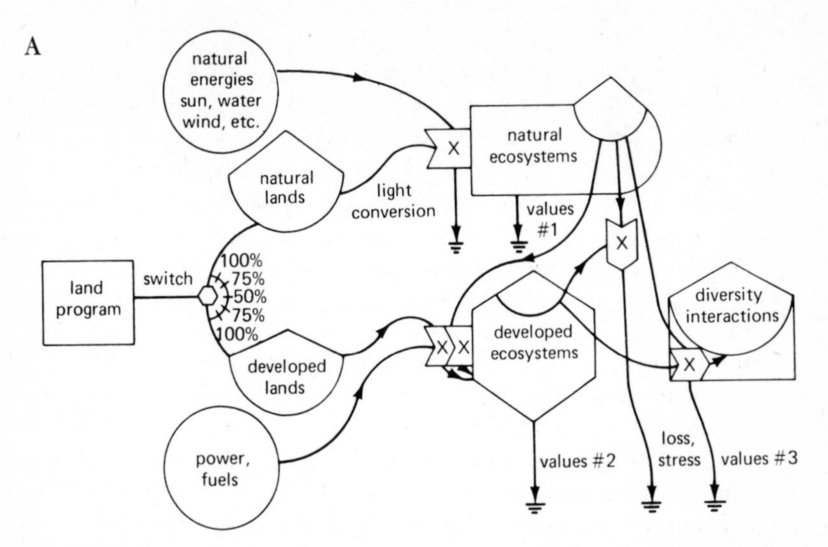

Fig. 8-6(A) A model for land management in which the proportion of natural and developed lands can be varied in order to determine the optimum balance in terms of the value of the total environment. (B) Performance curves based on the use of the model for a hypothetical region with extensive urban development. (After Odum and Odum, 1972.)

B

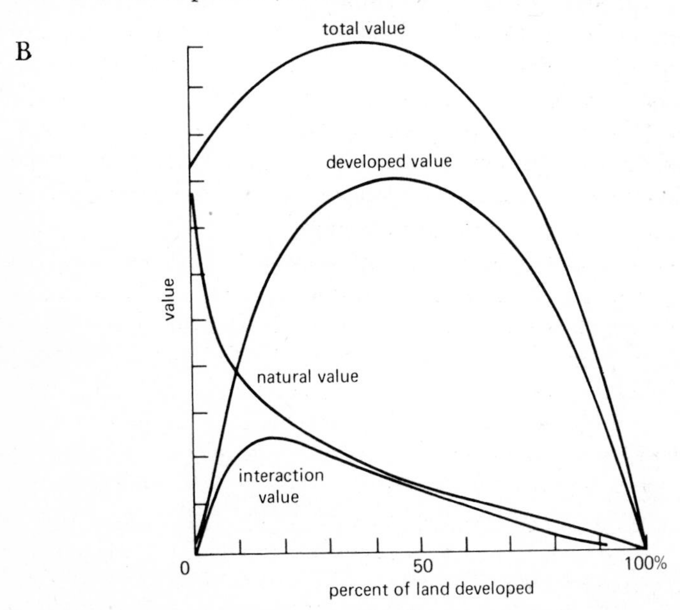

ment. Such curves are shown in Figure 8-6B for a hypothetical region containing one or more large cities. In this case the optimum plateau comes at about 40 percent developed lands; anything over 50 percent brings on a rapid decline in the total value. In regions with unfavorable climates or limited resources the optimum would likely be shifted to the left (less development desirable), while fertile regions could perhaps benefit from more intense development. For more on this, see H. T. Odum (1971) and Odum and Odum (1972).

Determining the optimum proportion of different kinds of landscapes is, of course, only *the first and very elementary step in comprehensive planning.* Spatial patterns are equally important, that is, where best to locate power plants, industrial parks, agricultural fields, and natural areas so as to achieve the most favorable interaction. As discussed in Chapter 6, a mixture of productive (developmental) and protective (climax) natural and seminatural communities is highly desirable.

Social problems, especially those relating to inequities, economic status, and social justice have overriding importance, especially where population density is high and industrial development intense. The impact of planning options must be determined not only with regard to the whole (that is, per capita), but also with regard to different social, racial, and economic groups which may be affected very differently by a given planning proposal. Thus, a highway may be projected as a net benefit for an urban region as a whole, but if it has a serious negative impact on a residential neighborhood, then it may, in fact, not be a positive benefit for the whole if the cost of social disorder cancels the projected economic gain.

Finally, the legal, economic, and political procedures necessary to implement ecosystem management cannot possibly be spelled out until there is a better scientific basis for comprehensive planning, some reliable way to determine carrying capacity, and an overwhelming public "mood for decision." The research and education tasks ahead are formidable! I prescribe to the notion that mankind can do almost anything he puts his mind to *provided he does not break natural laws in the process.* This book has been dedicated to outlining the natural laws within which we must work.

SUGGESTED READINGS

References cited

Barducci, T. B. 1972. Ecological consequences of pesticides used for the control of cotton insects in Canete Valley, Peru. In *The Careless Technology,* eds. M. T. Farvar and J. T. Milton, pp. 423–438. Garden City, New York: Natural History Press.

Barlowe, R. 1965. *Land Resource Economics.* Englewood Cliffs, New Jersey: Prentice-Hall.

Carson, Rachel. 1962. *Silent Spring.* Boston: Houghton-Mifflin.

Cloud, Preston, ed. 1969. *Resources and Man.* San Francisco: W. H. Freeman.

Gosselink, J. G.; E. P. Odum; and R. M. Pope. 1974. *The Value of the Tidal Marsh.* LSU-SG-74-03. Center for Wetlands Research, Baton Rouge, Louisiana.

Hardin, Garrett. 1968. The tragedy of the commons. *Science.* 162:1243–1248.

Heller, Alfred, Ed. 1972. *The California Tomorrow Plan.* Los Altos, California: William Kaufmann.

Inglis, D. R. 1973. *Nuclear Energy—Its Physics and Its Social Challenge.* Reading, Massachusetts: Addison-Wesley.

Lave, Lester B. and E. P. Seskin. 1970. Air pollution and human health. *Science.* 169:723–733.

Moncrief, Lewis W. 1970. The cultural basis for our environmental crisis. *Science.* 70:508–512.

Mumford, Lewis. 1967. Quality in the control of quantity. In *Natural Resources: Quality and Quantity,* eds. Ciracy; Wantrup; and Parsons, p. 7–18, Berkeley: Univ. California Press.

Odum, E. P. 1970. The attitude revolution. In *Crisis of Survival,* pp. 9–15. Glenview, Illinois: Scott, Foresman.

Szego, G. C. 1973. Future fuels—better energy management. In: *The Energy Crisis Symposium.* Atlanta: Western Electric Co.

Odum, E. P. and H. T. Odum. 1972. Natural areas as necessary components of man's total environment. *Trans. N.A. Wildl. and Nat. Res. Conf.* 37:178–189. Wildlife Management Inst., Washington.

Odum, H. T. 1971. *Environment, Power and Society.* New York: John Wiley & Sons.

Odum, Howard W. 1936. *Southern Regions of the United States.* Chapel Hill: Univ. North Carolina Press.

Odum, Howard W. and H. E. Moore. 1938. *American Regionalism.* New York: Henry Holt.

Spilhaus, A., ed. 1966. *Waste Management and Control.* See National Academy landmark reports.

Steinhart, J. S. and C. E. Steinhart. 1974. Energy use in the U.S. food system. *Science.* 184:307–316.

White, Lynn. 1967. The historical roots of our ecological crisis. *Science.* 155:1203–1207.

Williams, Carroll M. 1967. Third-generation pesticides. *Sci. Amer.* 217(1): 13–17.

The National Academy of Science Landmark reports

Environmental Quality and Social Behavior. Strategies for Research. 1973. A report by a panel of social and behavioral scientists. Nat. Acad. Sci., Washington.

Genetic Vulnerability of Major Crops. 1972. Nat. Acad. Sci., Washington. (Conclusion: "Most major crops impressively uniform and impressively vulnerable.")

Rapid Population Growth. 1971. Roger Revelle, ed. Published in book form by Johns Hopkins Press, Baltimore. (Conclusion: No economic advantage of rapid growth and many dangers.)

Resources and Man. (See citation, Cloud, 1969.)

The Earth and Human Affairs. 1972. Published in book form by Canfield Press, San Francisco. (A panel of geologists looks at man's impact on the biosphere.)

Waste Management and Control, ed. A. Spilhaus 1966. Publ. 1400. Nat. Acad. Sci., Washington.

Contrasting views on the technological society

Ellul, Jacques. 1964. *The Technological Society.* New York: Alfred A. Knopf. (Emphasizes the dehumanizing and social disordering aspects of technology and concludes that a dictatorship is only form of government capable of dealing with a high-energy, technological society.)

Spilhaus, A. 1972. Ecolibrium. *Science.* 175:711–715. (New technology and planned cities are capable of solving current crises and establishing once and for all a balance between mankind and environment.)

See also Starr and Rudman, *Science.* 182:258–364 for a similar view.

Contrasting views on the breeder reactor

Novick, Sheldon, 1974. Nuclear breeders. *Environment.* 16:6–15. (Critical of the Atomic Energy Commission's crash program which, according to the analysis, "overstates benefits, ignores hazards and fails to take into account alternatives including programs of other federal agencies.")

Weinberg, A. M. and R. P. Hammond. 1970. Limits to the use of energy. *Sci. Amer.* 58:412–420. (Authors marshal evidence that the breeder reactor is safe. They believe limits to population set by energy are very large; and estimate that 20 billion people can be supported if breeder reactor is developed or if controlled fusion becomes feasible. They point out, however, that such a dense world population would not be desirable for reasons other than energy needs. See also: Weinberg. *Science.* 177:27–34 and *Bull. Atomic Sci.* June 1970.)

Contrasting views on human population problem

Coale, A. J. 1970. Man and his environment. *Science.* 170:132–136. (A director of a Population Research Office argues that economic factors are more important than population growth in threatening the quality of life.)

Ehrlich, Paul R. and J. P. Holden. 1971. Impact of population growth. *Science.* 171:1212–1217. (A biologist and a physicist argue that rapid population growth is greatest threat to quality of life.)

The optimum population

Odum, Eugene P. 1970. Optimum population and environment: A Georgian microcosm. *Current History*. 58:355–359;365.

Singer, S. F., ed. 1971. *Is There an Optimum Level of Population?* New York: McGraw-Hill.

Taylor, L. R. ed. 1970. *The Optimum Population for Britain*. New York: Academic Press.

Ultsch, Gordon R. 1973. Man in balance with the environment: pollution and world population. *Science*. 185.13–19. (Suggests environmental pollution be used as an indicator of carrying capacity (*K*); when pollution is increasing the optimum level has been exceeded.)

Resources

Brooks, D. B. and P. W. Andrews. 1974. Mineral resources, economic growth, and world population. *Science*. 185:13–19. (Suggests "we are running out not of mineral resources but of ways to avoid ill effects of high rates of exploitation.")

Gaugh, W. C. and B. J. Eastlund. 1971. The prospects of fusion power. *Sci. Amer*. 224(2):50–64.

Goldman, Charles R.; James McEvoy, III.; and Peter J. Richerson, eds. 1973 *Environmental Quality and Water Development*. San Francisco: W. H. Freeman.

Hammond, Allen L. 1972. Photovoltaic cells; direct conversion of solar energy. *Science*. 178:732–733. (Although large amounts of electricity from solar cells is not likely for the near future, the potential is large.)

Makhijani, A. B. and A. J. Lichtenberg. 1972. Energy and well-being (on energy use in the U.S.). *Environment*. 14:8–13. (Although standard of living is correlated with per capita energy consumption, some countries have a high standard of living with less per capita energy use than the United States.)

McHale, John. 1970. *The Ecological Context*. New York: George Braziller. (A resource inventory approach to an overview of man's problems.)

McKelvey, V. E. 1972. Mineral resource estimates and public policy. *Amer. Sci*. 60(1):32–40.

Starr, Chauncey. 1971. Energy and power, and ten other articles in the September, 1971, issue of *Sci. Amer*. 229–3. Also published in book form by W. H. Freeman, San Francisco.

Wharton, C. 1969. The green revolution: cornocopia or Pandora's box? *Foreign Affairs*. 47:464–476.

See also "food for man" references, Chapt. 3.

Pollution

Benarde, M. A. 1970. *Our Precarious Habitat*. New York: Norton. (An integrated approach to understanding man's effect on his environment.)

Brodine, Virginia. 1971. *Air Pollution*. New York: Harcourt Brace Jovanovich. (Readable account with chapter on "reducing the burden.")

Clark, J. R. 1969. Thermal pollution and aquatic life. *Sci. Amer.* 220(3): 18–24.

Edwards, C. A. 1969. Soil pollutants and soil animals. *Sci. Amer.* 220(4): 88–93.

Goldwater, Leonard J. 1971. Mercury in the environment. *Sci. Amer.* 224(5):15–21.

Haagen-Smit, A. J. The control of air pollution. *Sci. Amer.* 209(1):24–29.

Hammond, A. L. and T. M. Maugh. 1974. Stratospheric pollution: multiple threats to earth's ozone. *Science*. 186: 335–338 ("it is increasingly clear that a wide range of human activities have the capability of disrupting the delicate photochemical balance on which the earth's ozone buffer, and perhaps life itself, depends." See also: Cutchis, 1974. Suggested Readings, Chapt. 3).

Kardos, L. T. 1970. A new prospect (on constructive uses of treated sewage). *Environment*. 12:10–21. (See also: Publ. 85, *Amer. Assoc. Adv. Sci.*, p. 241; *J. Soil & Water Cons.* 23.164; and *Water Research*. Vol. 2, New York: Pergamon Press, p. 371.

Ridker, Ronald. 1972. Population and pollution in the United States. *Science*. 176:1085–1090.

Stoker, H. S. and S. L. Seager. 1972. *Environmental Chemistry of Air and Water Pollution*. Glenview, Illinois: Scott, Foresman and Co. (A non-technical treatment with emphasis on specific pollutants such as CO, NO_2, SO_2, mercury, and so on.)

ven den Bosch, R. 1970. Pesticides: prescribing for the ecosystem. *Environment*. 12:117–119. See also "Cost of poisons," 14:12–14.

Woodwell, G. M. 1970. Effects of pollution on the structure and physiology of ecosystems. *Science*. 168:429–433.

Economics, technology, law, etc.

Anderson, Frederick R. 1973. *NEPA in the Courts*(*Resources for the Future*) distributed by the John Hopkins Press, Baltiomre. (A legal analysis of the National Environmental Policy Act, the single, most important federal law calling for study of the effect of proposed government actions on the environment.

Bigham, D.A. 1973. *The Law and Administration Relating to Protection of the Environment*. London: Oyez Publising Limited. (A survey treatise on environmental regulation in the United Kingdom illustrating how such controls are effected in another legal system.)

Boulding, Kenneth. 1964. The entropy trap. Chapter VII. In *The Meaning of the Twentieth Century—The Great Transition*. New York: Harper & Row.

Brooks, Harvey and Raymond Bowers. 1970. The assessment of technology. *Sci. Amer.* 222(2):13–20.

Curlin, J. W. 1971. Legal systems and ecosystems. *Student Law Jour.* January 1971: 4–8.

Daly, Herman E., ed. 1973. *Toward a Steady-State Economy.* San Francisco: W. H. Freeman.

Farvar, M. T. and J. P. Milton, eds. 1972. *The Careless Technology; Ecology and International Development.* Garden City, New York: Natural History Press. (Documents specific cases where large dams, irrigation projects, industrialized agriculture, and other technologies introduced into undeveloped countries have caused severe environmental and social problems.)

Fife, D. 1971. Killing the goose (on the profitability of destroying a natural resource). *Environment.* April 1971.

Fisher, A. C.; J. V. Krutilla; and C. J. Cicchetti. 1972. The economics of environmental preservation: a theoretical and empirical analysis. *Amer. Econ. Rev.* 62:605–619. (Applies mathematical model to cost and benefit analysis proposed Hell's Canyon Dam. The authors are leaders of a "new breed" of resource economists; watch for their future articles and books!)

Gregory, Derek P. 1973. The hydrogen economy. *Sci. Amer.* 228(1):13–21.

Kershaw, David N. 1972. A negative-income-tax experiment. *Sci. Amer.* 227(4):19–25.

McHugh, J. L. 1970. Economists on resource management (a review). *Science.* 168:737–739. (Regulating the catcher is better than regulating the catch in terms of both economics and sustained yield in commercial fisheries.)

Mishan, E. J. 1970. *Technology and Growth; the Price We Pay.* New York: Praeger. (Discusses some economic costs for growth that are not included in Chamber of Commerce promotional material!)

Ridker, R. G. 1973. To grow or not to grow: that's not the question. *Science.* 182:1315–1318. (The question is how to deal with consequences of growth as it occurs (not later) so as to provide negative feedback that will keep growth in bounds.)

Sax, Joseph. 1971. *Defending the Environment.* New York: Alfred A. Knopf. (A legal analysis prepared for non-lawyers on what lawyers and the courts can do in environmental disputes.)

Schurr, S. H., ed. 1972. *Energy, Economic Growth and the Environment.* Baltimore: Johns Hopkins Univ. Press. (Nine authors with contrasting viewpoints.)

Slesser, Malcolm. 1974. Energy analysis in technology assessment. *Tech. Assessment.* 2(3):201–208. London: Gordon and Breach. (Documents energy cost of making automobiles and producing food.)

See also references on economics, Chapt. 3.

Comprehensive planning and management

Altshuler, Alan. 1965. *The City Planning Process: A Political Analysis.* Ithaca, N.Y.: Cornell Univ. Press. (Political realities that have to be considered in any planning effort.)

Flawn, Peter T. 1970. *Environmental Geology. Conservation, Land-Use and Resource Management.* New York: Harper & Row. (A geologist documents the theme that shifting priorities and economic conditions make current methods of cost-benefit ratios for proposed earth alterations completely inadequate.)

Holling, C. S. and M. A. Goldberg. 1971. Ecology and planning. *Amer. Inst. Plan.* July 1971: 221–230.

Lee, K. N. 1973. Options for environmental policy. *Science.* 182:911–912. (A review of three books on joining economics and environment to form workable policy for the future.)

McAllister, D. M., ed. 1973. *Environment: A New Focus for Land-Use Planning.* National Science Foundation, Washington, D.C.

Meadows, D. L. and D. H. Meadows, ed. 1973. *Towards Global Equilibrium: Collected Papers.* Cambridge, Massachusetts: Wright-Allen Press. (A sequel to *Limits of Growth* showing what might be done to avoid the boom and bust predicted by the earlier study.)

Miller, G. T. 1972. *Replenish the Earth. A Primer in Human Ecology.* Belmont, California: Wadworth. (Interestingly written by a chemist. Nontechnical account of environmental science with emphasis on human population growth, thermodynamic principles, global food, and pollution crises. The approach is one of "cautious optimism" with a chapter on "what we must do.")

Milsum, J. H. 1969. Technosphere, biosphere and sociosphere: an approach to their systems modeling and optimization. *Ekistics.* 27:171–177.

Platt, John. 1973. Movement for survival. *Science.* 180:580–582. (A review of five books suggesting constructive action, as contrasted with mere rhetoric and polemics, for global analysis to implement a "world survival movement" for mankind.)

Reilly, W. K., ed. 1973. *The Use of Land: a Citizen's Policy Guide to Urban Growth.* New York: Thomas Crowell. (The Rockefeller Brothers report.)

Saarinen, Eliel. 1943. *The City, Its Growth, Its Decay, Its Future.* Cambridge, Massachusetts: M.I.T. Press.

Strong, A. L. 1971. *Planned Urban Environments.* Baltimore: Johns Hopkins press. (European experience suggests that only by a major commitment to national environmental planning can our urban crisis be resolved—a recognition by social scientists and urban planners of essential links between urban and rural environment.)

Walsh, John. 1971. Vermont: a small state faces up to a dilemma over development. *Science.* 173:895–987. (Striving to plan for development without destruction of landscape and to resist outside pressure that would force growth beyond the optimum.)

Watt, Kenneth. 1974. *The Titanic Effect: Planning for the Unthinkable.* Stamford, Connecticut: Sinauer. (Warns against the "it can't happen here" complacency; planning must include provisions for the unexpected.)

Appendix 1

APPENDIX 1. Energy Units (Source: *Handbook of Chemistry and Physics.* Cleveland, Ohio: Chem. Rubber Publ. Co.)

A—*Units of potential energy*

calorie or gram-calorie (cal or gcal)—heat energy required to raise 1 cubic centimeter of water 1 degree centigrade (at 15°C)
kilocalorie or kilogram-calorie (kcal)—heat energy to raise 1 liter of water 1 degree centigrade (at 15°C) = 1000 calories
British Thermal Unit (BTU)—heat energy to raise 1 pound of water 1 degree Fahrenheit
joule—work energy to raise one kilogram to height of 10 centimeters (or one pound to approximately 9 inches) = 0.1 Kilogram-meters
foot-pound—work energy to raise one pound, one foot = 0.138 Kilogram-meters.

Conversions:

1 calorie = 4.18 joules
1 kcal = 1000 cals = 3.97 (about 4) BTU = 4185 joules
1 BTU = 252 cal = 0.252 (about ¼) kcal
1 joule = 0.74 foot-pounds = 0.239 cal
1 foot-pound = 1.36 joules = 0.324 cal

B—*Units of Power (energy-time units)*

watt (w) (the standard international unit of power) = 1 joule per second = 0.239 cal per second
kilowatt-hour (kwhr) (the standard unit of electric power) = 1000 watts per hour = 3.6×10^9 watts
horsepower (hp) = 550 foot-pounds per second

Conversions

1 watt = 0.239 cal (per second)
1 kilowatt-hour = 860 kcal or 3413 BTU (per hour)
1 horsepower = 746 watts = 178 cal (per second)
Note: In this book kcal per day or year is used for comparison of natural and man-made power flow

C—*Energy content of some familiar quantities (in round figures)*

1 gram carbohydrates = 4 kcal
1 gram protein = 5 kcal
1 gram fat = 9 kcal
1 gram average plant biomass = 2 kcal/gm wet wt.; 4.5 kcal/gm ash-free dry weight
1 gram average animal biomass = 2.5 kcal/gm wt.; 5.5 kcal/gm ash-free dry weight
1 gram coal = 7.0 kcal
1 pound of coal = 3,200 kcal
1 gram gasoline = 11.5 kcal
1 gallon of gasoline = 32,000 kcal

Appendix 2

APPENDIX 2. Calculation of the Simpson and Shannon diversity indices (see Chapter 2).

If we let the letter "*N*" represent some specified importance value (number, biomass, energy flow, and the like) and designate *n* sub *i* (n_i) as the importance value for each component (species, for example) and capital *N* as the total of importance values—then we can say that the ratio n_i/N is the probability function, p_i, for each of the parts of the whole. Using these symbols we can now write the formulas for two of the most widely used diversity indices, the Simpson index which we will designate as *D*, and the Shannon index which we designate as *H*, as follows:

$$D = \Sigma \left(\frac{n_i}{N}\right)^2 \qquad H = -\Sigma \left(\frac{n_i}{N}\right) \log_e \left(\frac{n_i}{N}\right)$$

or or

$$D = \Sigma (p_i)^2 \qquad H = -\Sigma\, p_i \log_e p_i$$

The Sigma, (Σ) means "the sum of," which is to say that one computes the ratio for each component, squares it in the case of *D* and multiplies it by the log of the ratio for *H*, and then adds them all up to get the index number. Since logs of fractions are minus, a minus sign is inserted in the *H* formula to convert the sum to a positive number. All of this can be done quickly and easily with a pocket electronic calculator. If the calculator does not have a "log button" then it will be necessary also to have a table of natural logs. The two indices are named after the authors who have proposed the respective formulations. The Shannon index is of special interest because it is an approximation of a function originally proposed as a measure of information.

To inquire a bit more into what the indices tell us about the diversity of the system, let us compute these two diversity indices for the tall grass prairie from the data in Table 2-3. We suggest you do these calculations yourself, since once you get the knack of it, it is great fun to work out indices for all kinds of census data. Since the percentages given in Table 2-3 become probability, or relative abundance, ratios when converted to decimal fractions (move decimal two places to the left) can be shown as computations in table form as follows: S=number species having the indicated p_i value.

p_i	S	$p_i^2 \times S$	$p_i \log_e p_i \times S$
0.24	1	0.9576	0.3425
0.12	1	0.0144	0.2544
0.09	2	0.0162	0.4334
0.06	5	0.0180	0.8440
0.008	20	0.0013	0.7725
		$D=0.0175$	$H=2.6469$

Note that we have used an average value for the 20 rarer species; we could go back to the original source of this data and use actual field estimates, but this would prolong the calculations and would not appreciably alter the total index value since the p_i values are so small. The Simpson index, D, is an index of dominance since the maximum value, 1, is obtained when there is only 1 species (complete dominance), and values approaching zero are obtained when there are numerous species, each a very small fraction of the total (no dominance). When we are thinking in terms of diversity it is convenient to compute the reciprocal, $1-D$, so that the higher the value the greater the diversity. The Shannon index H is an index of diversity in that the higher the value the greater the diversity and the less the community is dominated by one or a few kinds. To compare the two directly it is necessary to scale the latter so that 1 is the maximum and zero the minimum. This can be done by dividing H by $\log_e$ of S (which is maximum possible index value for the number of kinds present). The Simpson index is already scaled 0–1, as previously indicated. Thus, for the stand of prairie vegetation we end up with the following:

$$1-D=0.8925 \qquad H/\log_e S=0.7861$$

In a general way we can now say that the plant diversity of the undisturbed prairie is 89 percent on the Simpson scale, and 78 percent on the Shannon scale, of the maximum possible diversity in a system of 29 species. Referring back to the computations we see that the 20 rare species contribute very little to the sum in the Simpson Index because squaring small fractions gives even smaller fractions. In contrast, the 20 rare species contribute substantially (almost 30%) to the Shannon index. Thus, the Simpson index is weighted in favor of the common species and the Shannon index in favor of the rare ones.

The diversity indices for the mite data and the grain field vegetation data as discussed in Chapter 2, are calculated in the same manner as outlined for prairie vegetation. Numbers of individuals and biomass in grams are the basis for importance in mites and vegetation, respectively.

For more on these and other diversity indices and their statistical treatment, see the following:

Hutcheson, Kermit. 1970. A test for comparing diversities based on the Shannon formula. *J. Theor. Biol.* 29:151-154.

Margalef, Ramon. 1963. On certain unifying principles in ecology. *Amer. Nat.* 97:357-374.

Odum, E. P. 1971. *Fundamentals of Ecology*, 3rd ed., pp. 148-154. Philadelphia: W. B. Saunders.

Shannon, C. E. and Weaver, W. 1963. The Mathematical Theory of Communication. Univ. Ill. Press, 117pp.

Simpson, E. H. 1949. Measurement of diversity. *Nature* 163:688

Appendix 3

Some Indices of Environmental Impact of an Individual U.S.A. Citizen in 1970

All figures are *per capita* (per person), and are *round figure* estimates for 1970. Water, energy, and timber figures are obtained by dividing total national consumption by 200×10^6 people (approx. population 1970).

	Physiological Need (Breathing, Drinking, and Eating)	*Total Consumption (Cultural needs, i.e., Including Household, Industrial, Agricultural, and other uses)*	*Ratio Total: Physiological*
Air:	300 cu. ft/day	5000 cu. ft/day	17:1
Water:			
Total used[a]	0.66 gal/day	2000 gal/day	3030:1
Total consumed[b]	0.3 gal/day	750 gal/day	2500:1
Energy:	1×10^6 Kcal/year ($= 4 \times 10^6$ BTU/year) (Food)[c]	87×10^6 Kcal/year ($= 345 \times 10^6$ BTU/year) (Food + all fuels)	87:1
Wood: (timber, paper, all other uses)	—	56 cu. ft/year[d] (a-prox. 2000 lb. or 1 ton dry wt/yr)	

[a] Total water withdrawn from surface streams and lakes plus underground sources.
[b] Amount water *not* returned to environment. Water returned to environment (subtract consumed from total) is polluted to varying degrees even where partially treated (secondary treatment) and thus has a direct impact on natural environment into which it is discharged.
[c] 2.0 to 2.5 acres farmland required to produce annual per capita food requirement with an American diet rich in meat. As little as 0.3 acres of intensively managed (fuel subsidized) land could support an adult human on a vegetarian diet.
[d] 56 cu. ft. equals 4 large trees 10 inches in diameter, or 14 small trees 6 inches in diameter. To grow 56 cu. ft. per year requires at least 0.3 acre *intensively managed forest land* (clear-cut pine plantation) or up to *1 acre forest land under multiple use* (timber production), a by-product of management for recreation, watershed protection, and other values).

Note: To estimate per capita impact for other countries divide figures in the total consumption column by ratio of that country's per capita energy consumption to U.S.A. per capita energy consumption.

Index

Numbers in **boldface type** indicate pages on which terms and concepts are most fully defined or explained. Place names and names of persons are not indexed (see reference lists at end of each chapter for the latter).